DRY-CURED MEAT PRODUCTS

DRY-CURED MEAT PRODUCTS

by

Fidel Toldrá, Ph.D.

INSTITUTO DE AGROQUÍMICA Y TECNOLOGÍA DE ALIMENTOS (CSIC)
PO BOX 73
46100 BURJASSOT (VALENCIA)
SPAIN

FOOD & NUTRITION PRESS, INC.
TRUMBULL, CONNECTICUT 06611 USA

FOOD & NUTRITION PRESS, INC.

6527 Main Street
Trumbull, Connecticut 06611 USA

Library of Congress Control Number: 2002103872

ISBN: 0-917678-54-0

DEDICATION

To my wife Milagro

To my children Fidel and Silvia

PREFACE

Dry-cured meat products, such as dry-cured ham and dry-fermented sausages, constitute one of the most representative traditional foods that have been produced and consumed throughout history by a diversity of cultures in different areas of the world. These meat products, which have a wide variety of flavors and textures, represent an important part of local economies, particular cultures and gastronomic heritages. Today, there is an important trend to enrich our sensory perceptions, and many consumers and meat industries around the world are getting more and more interested in dry-cured meat products.

This book presents the latest developments in dry-cured meat products, from raw materials and manufacture to the final products, and includes updated scientific and technological information, especially on the safety, quality and nutritional properties of these foods. The first chapter provides a historical perspective, and Chap. 2 provides information on the composition, organization and enzyme system of muscle. Chapters 3 and 4 present descriptions of the manufacturing processes, including old and new technologies and trends for acceleration. Chapter 5 covers fermentation aspects and the main characteristics of starter cultures used in the processing of dry-fermented sausages. Chapters 6 and 7 present information on the two important groups of biochemical reactions, proteolysis and lipolysis, respectively. Chapter 8 covers the important flavor development characteristic of dry-cured meat products. Nutritional properties are presented in Chap. 9. Chapter 10 presents the effects of raw materials and processing on the quality of these meat products, and Chap. 11 presents the main defects and methods for prevention. Safety and economic aspects are discussed in Chap. 12 and 13, respectively.

This book is written as a text for advanced undergraduate and graduate students. It will also be useful as a reference for basic/applied scientists involved in this field, and food technologists working in quality control and R&D departments in the meat industry.

FIDEL TOLDRÁ

ACKNOWLEDGMENTS

First of all, I wish to express my most sincere gratitude to my colleague and very good friend, Robert G. Cassens, for his helpful advice and for encouraging me to write this book. I am very grateful to him, because without his help, this book would not have been written.

Each chapter has been read and criticized by colleagues with expertise in the field who provided me with useful observations and interesting suggestions. In this regard, I wish to thank R.G. Cassens (Univ. of Wisconsin-Madison, U.S.), G. Monin (INRA, Clermont, France), D. Demeyer (Univ. of Ghent, Belgium), D.J. Troy (The National Food Centre, Dublin, Ireland), J.M. Monfort and J.A. García-Regueiro (CTC-IRTA, Girona, Spain), G. Gandemer (INRA, Lusignan, France), J. Ventanas (Univ. of Extremadura, Cáceres, Spain), P. Baldini (Stazione Sperimentale per l'Industria delle Conserve, Parma, Italy), J. Flores (IATA-CSIC, Valencia, Spain) and P. Roncalés (Univ. of Zaragoza, Spain).

I also wish to thank John O'Neil, Maureen Yash, and Jennifer Read of Food & Nutrition Press for editorial assistance and Mr. P. Rico who assisted me with the artwork.

CONTENTS

CHAPTER 1

INTRODUCTION: A HISTORICAL PERSPECTIVE

Historical Developments

One of the first concerns of the primitive man when he started to think was, undoubtedly, wondering what he would eat and when. Instinctively, hunters ate the fruits they found and the animals and fish they were able to catch.

The populations settled in fertile lands and had plenty of easy hunting. But, when these resources became exhausted or the climate changed, great migrations of the population took place as they looked for more favorable lands. Nevertheless, step by step, primitive man improved his mind and physical abilities and learned how to harvest and domesticate farm animals. Toward 8000–10000 B.C., agriculture and farming were initiated.

At that time, there were two main concerns for the primitive man: how to obtain the food for the future, and, in case of great success, how to preserve the excess of foods. This was the beginning of the need for preservation of excedentary foods in order to ensure subsistence for times of scarcity. Different techniques were successfully assayed and are still in use today, although with a highly advanced technology.

The cold storage of foods could be used in cool areas like northern Europe, and the use of drying could be applied in areas with mild winters and nonexcessive rainfalls, like the Mediterranean. Other techniques like smoking and cooking appeared after the discovery and application of fire. On the other hand, the origin of salting as a preservative technique was lost in ancient times. Primitive man rubbed pieces of meat and fish with marine salt, which contained nitrate as an impurity. Good preservation was obtained. Salting is still in use in different areas with low evolution. Today, camel caravans still cross the Sahara desert following millenary routes and carrying salt as a precious good. Until recently, salt commercialization has been the object of protectionist laws and special taxes.

The first harvests were distributed through different cultural and climatic zones. In general, crops consisted of wheat, corn, barley, rice, and fruits. The first farming animals were cows, sheep, pigs, rabbits and poultry. Along with milk and eggs, their skin, bones, wool and feathers were highly valued. With successive migrations, the populations became homogenized and expanded their food sources. However, some populations had preferences toward one animal, such as the American Indians who had many troubles when the buffalo were exterminated. Other populations had severe cultural or religious restrictions that forbid the consumption of pork and its derivatives, such as Jews, Arabs and some populations in Asia and Africa.

The production and consumption of dry-cured meats probably originated in southern European countries around the Mediterranean sea because its particular climate allows natural drying and ripening. On the other hand, the use of smoke was applied in northern and colder areas where the climate did not allow for natural drying.

There are many historical references to pork and its dry-cured derivatives. In one of the numerous Sumerian tablets, written in the cuneiform language by the year 2000 B.C., there are numerous references to pork as an important food source at that time in Sumer (Kramer 1965): "He was with no resources and then he had to slaughter the pig". "The butcher slaughters the pig saying: but, is it necessary to cry? It is the same way followed by your father and grandfather and now you are following it. And, however, you are still crying!"

According to Egyptian mythology, Seth was introduced in the eye of Horus in the form of a black pig, blinding him. Perhaps this was the beginning of the rejection of pork in some Afro-Asian populations (Max Müller 1996). Almost all Egyptian gods were represented by animals (bulls, cows, birds, etc.).

Dry-fermented sausages were already known by the ancient Romans and Greeks, in coincidence with the great expansion of pig use throughout Europe. Full details of pig slaughtering and the cooking of its products were described ca. 900 B.C. by Homer in Chap. XIV, vs 414-533, of the famous *Odyssey* (Homer 1993). The pigpen of Ulysses was described in Chap. XIV, vs 1-412, of the same book. In the first century, Petronio described the famous banquet of Trimalción in his book *El satiricón* (Petronio 1965). One of the dishes in the banquet consisted of products from wild pig. The effect of feed on the pork meat quality, expressed as the excellent influence of acorn, was already emphasized: "Have you seen what selected acorn ate that wild pig!" Romans, who tried to keep plenty of foods, found salted meats to be an easy food to store and transport. The Latin expression "salsicia" can be the etymological precedent of the Spanish "salchichón" or the French "saucisson." In a similar way, the Latin word "salumen," meaning a group of salted products, could be the origin of the Italian salami. Other authors attribute its origin to the city of Salamis in Cyprus (Zeuthen 1995). In his *De Re Agricola,* Catón described some recipes of salted meat products that are nowadays still consumed in certain Mediterranean areas (Pineda 1989). In the fifth century, the French butchers known as "charcutiers" already prepared different pork derivatives. In the twelfth century, the Saint Antoine l'Abbaye practiced innovative medicine that expanded throughout Europe using pork meat and lard. Pork meat products were widespread in consumption during the Middle Ages. Therefore, a lot of pork meats and meat products are mentioned in the famous book *El Quijote* written by Miguel de Cervantes in 1604. In this book there is a special mention to Dulcinea del Toboso who had the better hands for pork salting.

The modern dry-fermented sausage was apparently invented around 1730 in Italy, being later adopted around 1780 by the German countries after the stay in Italy of a German butcher named Butleb (Leistner 1992). The climatic conditions of the production area had a strong influence on the manufacture procedure. Areas with high humidity would need mechanical drying facilities. The climatic variations are even reflected in the names of some sausages. For instance, the winter salami, originally from the north of Italy, had to be produced in Hungary during the winter months for simulation of the adequate processing conditions. The summer sausage was produced in the summer and was heated for safety reasons (Zeuthen 1995).

The good experience and practice in manufacturing dry-cured meat products in Europe naturally went to America with the settlers. Pork was cured in New England in the 1600s for use in the summer (Kemp 1982). The production and consumption of cured meat products extended to the southeast of the United States. Different recipes (salt, salt and sugar, pepper, other spices, etc.) were used by curers, some of whom had ice houses and smoke houses on their farms. The Meat Inspection Act of 1906 allowed the use of nitrate. During the early 1920s, experiments were done on the use of nitrite. In 1925, rules were enacted about the use of nitrite for curing. The enactment of the Wholesome Meat Act took place in 1967. It became mandatory that curing procedures be approved either to comply with federal or state regulations and thus some guidelines had to be established as recommendations for processors (Kemp 1982). A lot of research was then carried out on the optimization of the production and quality of these products. Similarly, important research efforts have been developed in Europe in the last few decades.

Classification of Cured Meats

The term cured meat is used for a great number of meat products, although its meaning may vary depending on the kind of product and country of origin. Traditionally, the term curing generally means the use of a curing salt (sodium chloride and nitrate/nitrite) which generates or produces characteristic color and flavor in the product. Thus, curing is a broad term applied to many different meat products. Basically, cured-meat products can be arranged into two main groups based on their respective processes (Flores and Toldrá 1993): dry-curing and wet or pickle-curing, as illustrated schematically in Fig. 1.1.

Dry-curing, as this book details, involves the use of a dry cure that consists of salt alone or combined with nitrate and/or nitrite, which is rubbed on the surface of an entire piece (i.e., ham or shoulder) or mixed with the mince (sausage). On a very minor scale, the cure may be dissolved in water to form a brine that penetrates into the product by soaking (i.e., dry-cured loins). In all of these cases, the additives are transported by diffusion through the moisture

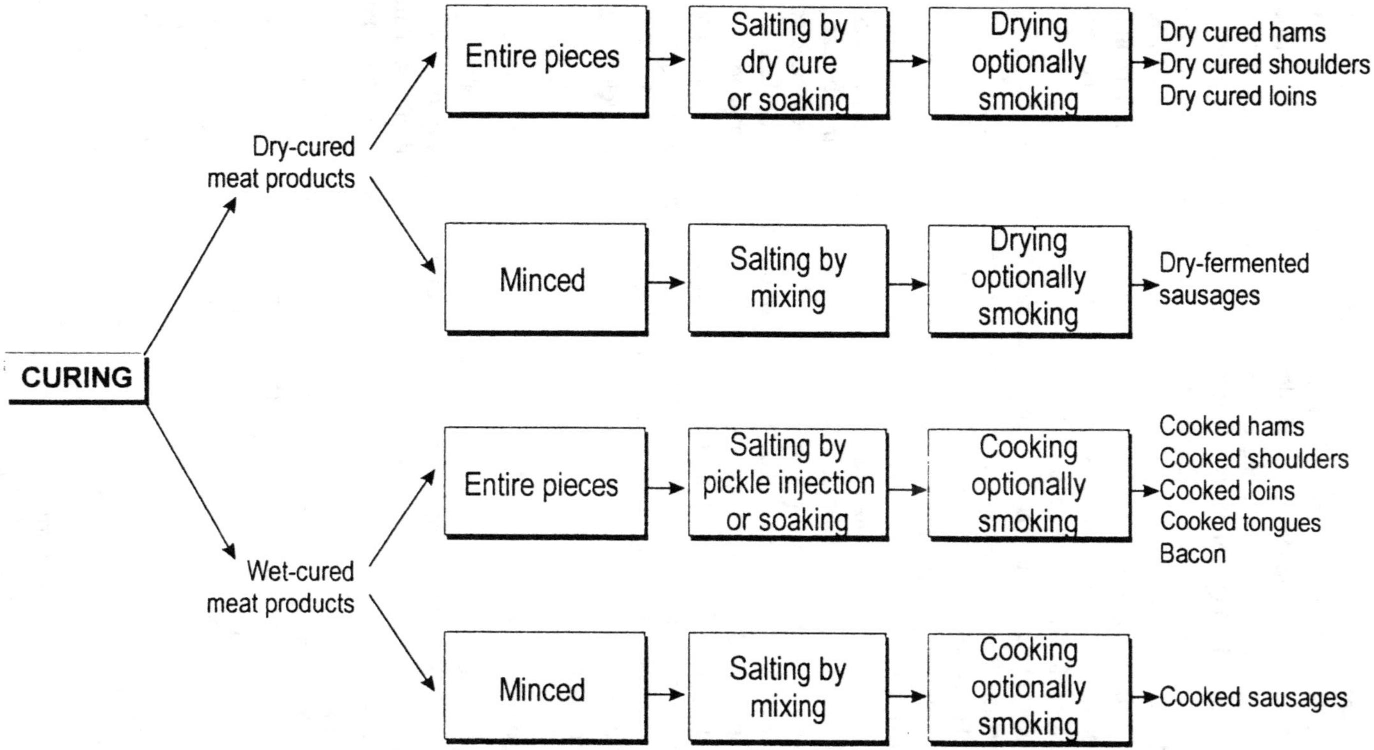

FIG. 1.1. DIAGRAM SHOWING THE CLASSIFICATION OF THE MOST-IMPORTANT CURED MEATS AND THE MAIN DIFFERENCES BETWEEN DRY AND WET CURING
(Adapted from Flores and Toldrá 1993)

of the meat. Finally, the products that can be optionally smoked are aged/ripened and dried for several weeks, months or even years.

On the other hand, wet curing consists of the pickle injection of the cure into the piece either by pumping the brine through the arterial system, stitch pumping using a needle with several openings along its length, or multiple needle injection pumping (Townsend and Olson 1987). Alternatively, the entire piece can be soaked into the curing brine (i.e., tongue), or the mince can be mixed with the cure to form a paste (i.e., cooked sausage). The brine serves as a vehicle of penetration into the meat products that are finally heat treated and, optionally, smoked simultaneously with the heat treatment. An exception is bacon, which is dried for a short time and/or smoked. Some typical wet-cured meat products are frankfurters and mortadella.

REFERENCES

FLORES, J. and TOLDRÁ, F. 1993. Curing: Processes and applications. *In* Encyclopedia of Food Science, Food Technology and Nutrition, R. Macrae, R. Robinson, M. Sadle and G. Fullerlove (eds.). Academic Press, London.

HOMERO 1993. La Odisea. Translation into Spanish by L. Riba from The Odissey.La Magrana, Barcelona, Spain.

KEMP, J.D. 1982. Nitrates and nitrites in country cured products. Proc. Reciprocal Meat Conf. *35,* 53-55.

KRAMER, S.M. 1965. From the Tablets of Sumer. The Falcon's Wing Press, Indian Hills, CO.

LEISTNER, F. 1992. The essentials of producing stable and safe raw fermented sausages. *In* New Technologies for Meat and Meat Products, J.M. Smulders, F. Toldrá, J. Flores and M. Prieto (eds.). Audet, Nijmegen, The Netherlands.

MAX MÜLLER, F. 1990. Egyptian mythology. Ed. Comunicación SA, Barcelona, Spain.

PETRONIO, 1965. El satiricón. Translation into Spanish from The satiricon. Ed. Plaza & Janés, Barcelona, Spain.

PINEDA, J.M. 1989. Influencia de la materia prima en la calidad del jamón curado. *In* Avances en la Tecnología del Jamón Curado, J. Flores, S. Bermell and F. Toldrá (eds.). Taroncher, Valencia, Spain.

TOWNSEND, W.E. and OLSON, D.G. 1987. Cured meats and cured meat products processing. *In* The Science of Meat and Meat Products, J.F. Price and B.S. Schweigert (eds.). 3rd ed., Food & Nutrition Press, Inc., Trumbull, CT.

ZEUTHEN, P. 1995. Historical aspects of meat fermentations. *In* Fermented Meats, G. Campbell-Platt and P.E. Cook (eds.). Blackie Academic & Professional, London.

CHAPTER 2

DESCRIPTION OF MAIN MUSCLE CHARACTERISTICS

Muscle Structure

There are three types of muscles: skeletal, smooth and cardiac. Skeletal muscle is the most important muscle in meat products because it comprises a high percentage of total body weight. It is voluntary (the organism can cause it to contract voluntarily), striated (cross striations are related with the contraction-relaxation mechanism) and multinucleated (several nuclei are located peripherally in the cell).

The muscle is divided into sections by the perimysium (thin connective tissue layers), and each section is divided into fibers that are individually wrapped by the endomysium (thin collagen fibers). Each muscle fiber is about 50 μm in diameter and up to several centimeters in length. It contains about 1,000 myofibrils, arranged in a parallel way, that are responsible for the contraction of the muscle. Myofibrils are composed of thick and thin filaments arranged in an array, giving rise to dark (A) and light (I) bands. The Z line and M line bisect each I and A band, respectively (Fig. 2.1). The sarcomere, which usually ranges from 2–3 μm in length, is the distance from Z line to the following Z line.

The thick and thin filaments are composed of proteins, which are responsible for muscle contraction and relaxation. The most important proteins are listed in Table 2.1. The scanning electron microscopy of pork skeletal muscle reveals details of the muscle fiber structure (Fig. 2.2). There are two main types of muscle fibers, red and white, and the appearance and properties of the muscle will vary according to the proportion of both types of fibers. So, differences in color of the cut surface of different muscles may be appreciated in pork and poultry. Red muscle is more aerobic in metabolism but contracts slower than white muscle. Additionally, red muscle has higher levels of myoglobin and lipids, lower levels of glycogen and is rich in oxidative enzymatic activity in opposition to the higher concentration of glycolytic enzymes in white muscle.

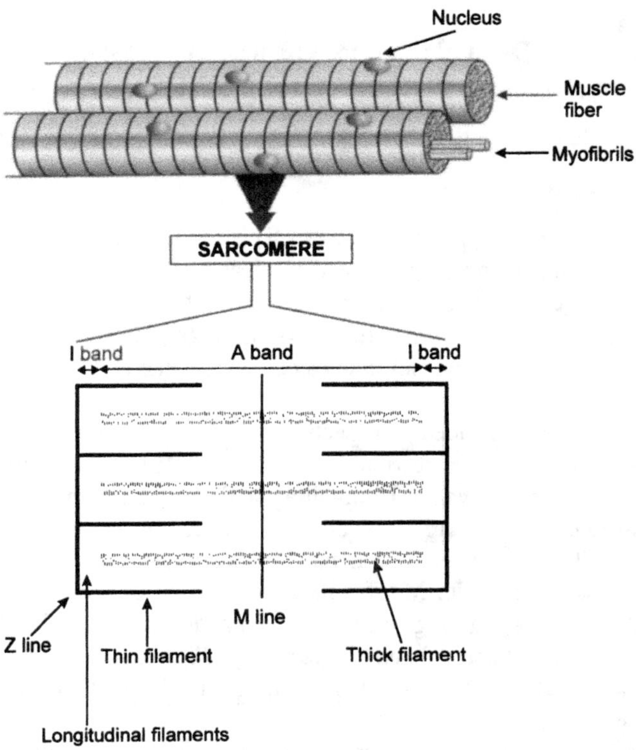

FIG. 2.1. A DIAGRAMMATIC REPRESENTATION SHOWING SKELETAL MUSCLE STRUCTURE FROM A MUSCLE FIBER TO MYOFIBRILS

TABLE 2.1.
MAIN CHARACTERISTICS OF MUSCLE MYOFIBRILLAR PROTEINS

Protein	Location	Role	Molecular mass (KDa)	% Total myofibrillar proteins
Myosin	Thick filaments	Contractile	520	43
Actin	Thin filaments	"	42	22
Tropomyosin	Thin filaments	"	68	5
Troponin	Thin filaments	"	69	5
Titin	Longitudinal filaments	Cytoskeletal	2,800	8
Nebulin	Thin filaments	"	600	3
C protein	Thick filaments	"	140	2
Actinin	Z line	"	200	2
M protein	M line	"	160	2
Desmin	Transversal filaments	"	55	<1

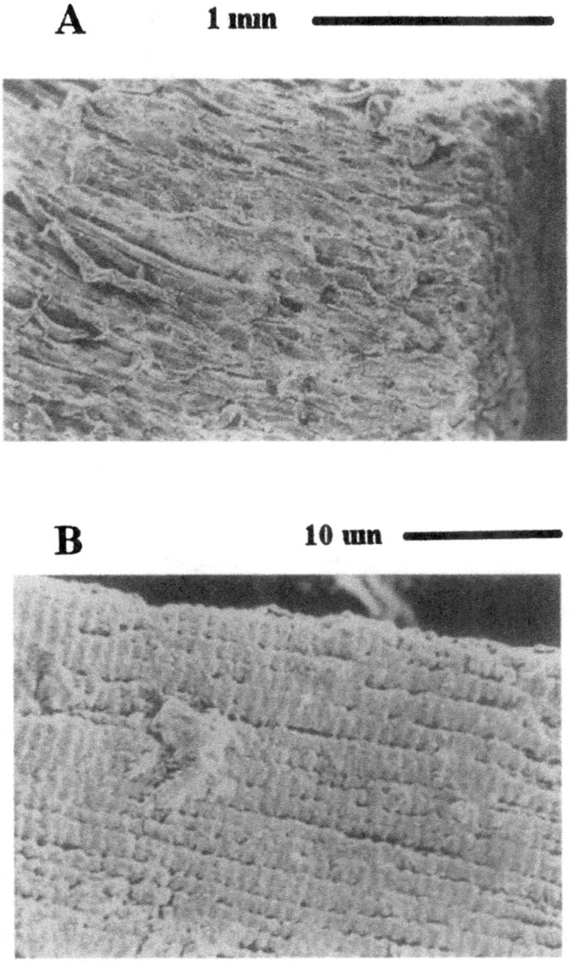

FIG. 2.2. SCANNING ELECTRON MICROGRAPHS OF PORCINE SKELETAL MUSCLE
(A) Muscle fibers with collagen of the endomysium surrounding each fiber.
(B) View of myofibrils showing the striations.
(Toldrá and Voyle 1988, unpublished)

MUSCLE COMPOSITION

Meat is basically composed of water, protein, lipid, minerals and trace amounts of carbohydrate. Lean muscle tissue contains approximately 74% water, 21% protein, 4% fat and 1% ash. These proportions may change depending on the amount of fattening and, especially if adipose tissue is included. The percentages of protein and water decrease when fat increases. An example of the typical composition of pork muscle is given in Table 2.2. A brief description of proteins and lipids as major components of meat is given below.

TABLE 2.2.
EXAMPLE OF COMPOSITION IN MUSCLE *SEMIMEMBRANOSUS* FROM (LANDRACE X LARGE WHITE) X LARGE WHITE 6-MONTH-OLD PIGS

	Content (g/100 g)		**Phospholipids (PL)**	**Content (% of total PL)**
Moisture	75.0		Phosphatidylethanolamine	22.76
Protein	21.9		Phosphatidylinositol	11.87
Total lipids	2.1		Phosphatidylcholine	59.18
Non-polar lipids	1.5		Cardiolipin	5.15
Phospholipids	0.584	} →	Phosphatidylserine	<1
Ash	0.9		Sphingomyelin	<1
Cholesterol	0.050			

(Adapted from Armero *et al.* 2001)

Muscle Proteins

Proteins are the major constituent of muscle and make up approximately 15–22%. Muscle proteins have a special importance during the conversion of muscle to meat and also during meat processing as they give rise to many compounds such as peptides, free amino acids, etc. There are three main groups of proteins in the muscle: myofibrillar, sarcoplasmic and stromal proteins.

Myofibrillar Proteins. These proteins are soluble in high ionic strength buffers. They are the main constituents of the structure of the myofibrils, and, in fact, myosin and actin provide the structural backbone of the myofibril. Tropomyosin and troponin are regulatory proteins associated with muscle contraction. The proteins in Z line serve as bridges between the thin filaments of adjacent sarcomeres. Titin and nebulin are two very large proteins, running in parallel to the long axis of the myofibril, that contribute to the longitudinal continuity and integrity of muscle cells (Robson *et al.* 1997). Desmin connects adjacent myofibrils at the level of Z line.

Sarcoplasmic Proteins. These proteins are water-soluble proteins, composing about 30–35% of the total protein in muscle. Sarcoplasmic proteins contain a high diversity of proteins, mainly metabolic enzymes (mitochondrial, lysosomal, microsomal, nucleus or free in the cytosol) and myoglobin. Some hemoglobin may also remain in the muscle although most blood is drained from muscle during bleeding. Myoglobin is the protein pigment responsible for the red meat color The amount of myoglobin depends on the fiber type and also varies according to the species. In general, the content in myoglobin increases with the age of the animal.

Stromal Proteins. Collagen is the basic protein-forming part of the connective tissue that surrounds the fibers and muscles (epimysium, perimysium and endomysium). Collagen provides strength and support to the muscle structure. It becomes tougher with age due to an increasing number and type of crosslinks. This is the reason why meat tenderness decreases in older animals. Elastin is found in lower amounts, usually in arterial walls.

Muscle and Adipose Tissue Lipids

Lipids in skeletal muscle make up about 1–13% of total muscle, depending on the degree of fattening and amount of adipose tissue. Main lipids are found intramuscularly, intermuscularly and in adipose tissue. Intramuscular lipids are mainly composed of triglycerides, stored in fat cells, and phospholipids, located in cell membranes. The amount of cholesterol in lean meat is around 50 mg/100 g. Intermuscular and adipose tissue lipids are mainly composed of triglycerides and small amounts of cholesterol (40–50 mg/100 g). There are two main groups of lipids in the muscle: nonpolar lipids, mostly triglycerides, and phospholipids.

Triglycerides. These are the major constituents of fat. The fatty acid content mainly depends on age, type of feed and environment. The composition of fat reflects the diet of the animal, especially in pork and poultry. In the case of ruminants, the nutrients are somehow standardized due to the action of the microbial population of the rumen. The properties of the fat will depend on the composition in fatty acids. In general, they tend to be sterified to saturated and monounsaturated fatty acids (Table 2.3). It will be soft (oily appearance) and prone to oxidation when there is a high percentage of polyunsaturated fatty acids like linoleic (typical of feeds rich in corn, for instance) and linolenic acids.

Phospholipids. These are present in minor amounts but have a strong importance for flavor development and oxidation in postmortem meat. Phospholipids have a relatively high proportion of polyunsaturated fatty acids in comparison to neutral lipids. Phosphatidylcholine (lecithin) and phosphatidylethanolamine are the major constituents. Phospholipids vary depending on the genetic type of the animal and anatomical location of the muscle. Therefore, the amount of phospholipids tends to be higher in red oxidative muscles than in white glycolytic muscles (Hernández *et al.* 1998).

TABLE 2.3.
EXAMPLE OF COMPOSITION IN MAJOR FATTY ACIDS OF NON-POLAR LIPIDS AND PHOSPHOLIPIDS IN THE MUSCLE *SEMIMEMBRANOSUS* FROM (LANDRACE X LARGE WHITE) X LARGE WHITE 6-MONTH-OLD PIGS

Fatty acids (FA)	Non-polar lipids	Phospholipids
C 14:0	1.41	0.23
C 16:0	22.80	20.24
C 18:0	10.01	15.25
C 16:1	3.36	0.86
C 18:1	47.62	17.00
C 20:1	0.78	0.18
C18:2	11.8	31.55
C 18:3	0.99	0.64
C 20:2	0.50	0.45
C 20:3	0.10	1.09
C 20:4	0.38	11.08
C 22:4	0.10	0.96
Total Saturated FA	34.37	36.13
Total Monounsaturated FA	51.87	18.10
Total Polyunsaturated FA	13.86	45.78

(Adapted from Armero *et al.* 2001)

The Muscle Enzyme System

The muscle contains a great variety of enzymes, responsible for most of the biochemical changes observed during the processing of meat and meat products. Some of the most important muscle enzymes are associated with protein breakdown (proteinases), the generation of small peptides (peptidases) or free amino acids (aminopeptidases and carboxypeptidases) and lipids breakdown (lipases). The location of the enzymes may also differ for each one. Some are located in lysosomes, and others are free in the cytosol, etc. A brief description of each group is given.

Lysosomal Proteinases. Lysosomes contain a great variety of enzymes including proteinases. The main lysosomal proteinases are cathepsins B, H and L, which are cysteine proteinases, and cathepsin D, which is an aspartate proteinase. These enzymes are small, in the range of 20–40 KDa (Table 2.4) and thus once free are able to penetrate into the myofibrillar structure. They are optimally active at acid (3.0–5.0 for cathepsin D), slightly acid (around 6.0 for cathepsins B and L) or even neutral (6.8 for cathepsin H) pH values and in the 30–40C range (Toldrá *et al.* 1990, 1991). The presence of these enzymes in skeletal muscle cells has been demonstrated by cytochemical, immunocytochemical or immunofluorescence techniques. The cysteine proteinases require a reducing environment to express their optimal activity. The anaerobic glycolysis in postmortem muscle generates the adequate environment for cathepsin activity (Etherington 1987).

There are numerous *in vitro* studies about the action of lysosomal proteinases on myofibrillar proteins. Cathepsins B, D, H and L have shown a good ability to degrade different myofibrillar proteins. Thus, Cathepsin D and L are especially active against myosin heavy chains, titin, M and C proteins, tropomyosin and troponins T and I (Matsakura *et al.* 1981; Zeece and Katoh 1989). Cathepsin L is extremely active in degrading both titin and nebulin. However, actin is degraded but at a slow rate. Cathepsin B is able to degrade myosin heavy chain and actin but has no effect on myosin light chains and troponin C (Schwartz and Bird 1977). Cathepsin H shows both endo and aminopeptidase activity being classified as an aminoendopeptidase. Although cathepsin H may degrade myosin, its main function seems to be as an aminopeptidase (Okitani *et al.* 1981).

Neutral Proteinases: Calpains. Calpains are a group of cystein endopeptidases located in the cytosol but, very especially, in the Z-disc region. A variety of names have been used in the scientific literature such as calcium-activated neutral proteinase, calcium-dependent protease and calcium-activated factor. Calpains I (μCANP) and II (mCANP) require 50–70 μM of Ca^{2+} and 1–5 mM of Ca^{2+}, respectively, for activation and show maximal activity at neutral pH (around 7.5), very poor activity below pH 6.0 or are probably ineffective at pH 5.5–6.0 (Etherington 1985). Calpains are heterodimers of 110 KDa composed of an 80 KDa catalytic subunit and a 30 KDa subunit of unknown function (Bond and Butler 1987). Calpains are able to degrade titin, nebulin, troponins T and I, tropomyosin, C-protein, filamin, desmin and vinculin resulting in large peptides but are not active against myosin, actin, α-actinin and troponin C (Goll *et al.* 1983; Koohmaraie 1994).

TABLE 2.4.
MAIN CHARACTERISTICS OF PORK MUSCLE ENDO-PROTEASES

Enzyme	EC number	Classification	Location	Molecular mass (KDa)	Optimal pH	Optimal Temp. (°C)
Cathepsin B	3.4.22.1.	Cystein	Lysosome	25	5.5-6.0	37
Cathepsin D	3.4.23.5.	Aspartyl	Lysosome	42	4.0	40
Cathepsin H	3.4.22.16.	Cystein	Lysosome	26	6.8	37
Cathepsin L	3.4.22.15.	Cystein	Lysosome	29	5.5-6.0	30
μ-calpain	3.4.22.17.	Cystein/metallo	Z-disc	110	7.5	25
m-calpain	3.4.22.17.	Cystein/metallo	Z-disc	110	7.5	25

Calpain I has a poor stability in postmortem muscle. It may be autolyzed in the presence of Ca^{2+} and even an excess of substrate, resulting in a loss of enzymatic activity that is highly temperature-dependent (Koohmaraie 1994). Calpain II seems to be more stable for a few weeks (Koohmaraie *et al.* 1987). This high unstability makes it unlikely that calpains might have a significant role in the dry-curing of hams. Furthermore, the acid pH values of dry-fermented sausages makes any calpain activity rather unlikely.

The existence of calpastatin, an endogenous inhibitor, regulates the activity of calpains in postmortem muscle. Pork muscle has the lowest calpastatin level when compared to other species (Ouali and Talmant 1990). Calpastatin is destroyed by autolysis in a few days post-slaughter (Koohmaraie *et al.* 1987).

Muscle Tripeptidylpeptidases. Tripeptidylpeptidases (TPP) are enzymes capable of hydrolyzing different tripeptides from the amino termini of peptides. TPP I is located in the lysosomes, has an optimal acid pH (4.0) and hydrolyzes tripeptides Gly-Pro-X, where X is an amino acid, preferentially of hydrophobic nature (Table 2.5). TPP II is a relatively big enzyme with a molecular mass higher than 1,000 KDa, an optimal neutral pH (6.5–7.5) and a wide substrate specificity, except when Pro is present on one of both sides of the hydrolyzed bond. Ala-Ala-Phe is a typical substrate optimally hydrolyzed.

Muscle Dipeptidylpeptidases. Dipeptidylpeptidases (DPP) are enzymes capable of hydrolyzing different dipeptide sequences from the amino termini of peptides. There are four different activities, named DPP I, DPP II, DPP III and DPP IV. DPP I and II are located in the lysosomes, DPP III is in the cytosol and DPP IV is linked to the plasm membrane. Their molecular mass is relatively high, between 100 and 200 KDa, and the optimal pH either acid (5.5 for DPP I and DPP II) or basic (7.5–8.0 for DPP III and IV). These four enzymes have been purified and fully characterized in porcine skeletal muscle (Sentandreu and Toldrá 1998, 2000, 2001a, 2001b).

DPP I is the most active in postmortem muscle and keeps an important percentage of activity at low temperatures so that it can be very active during salting and postsalting stages (Sentandreu and Toldrá 2001c). DPP have different substrate specificities. For instance, DPP I has a special preference to hydrolyze the dipeptides Ala-Arg and Gly-Arg, and DPP II and DPP IV for Gly-Pro and DPP III for Arg-Arg and Ala-Arg. In any case, these enzymes are able to hydrolyze other dipeptides with different sequences, although at a lower rate. These enzymes are therefore involved in the generation and accumulation of a wide range of dipeptides in postmortem meat and during the processing of meat products.

TABLE 2.5.
MAIN CHARACTERISTICS OF PORK MUSCLE PEPTIDYLPEPTIDASES

Enzyme	EC number	Classification	Location	Molecular mass (KDa)	Optimal pH	Optimal Temp. (°C)
DPP I	3.4.14.1	Cystein	Lysosome	200	5.5	45-55
DPP II	3.4.14.2.	Serin	Lysosome	116	5.5	65
DPP III	3.4.14.4.	Serin	Cytosol	80	8.0	40-45
DPP IV	3.4.14.5.	Serin	Membrane	150	7.5-8.0	45
TPP I	3.4.14.9.	Serin	Lysosome	55	4.0	37
TPP II	3.4.14.10.	Serin	Cytosol	1,000	6.5-7.5	30

Muscle Dipeptidases. Dipeptidases are another group of enzymes present in muscle although they are not yet fully purified and characterized. These enzymes catalyze the hydrolysis of dipeptides and depending on the preference for certain amino acids receive different names. For instance, glycylglycine dipeptidase is very specific for dipeptides containing glycine, cysteinylglycine dipeptidase is specific for the dipeptide Cys-Gly and arginin dipeptidase has special preference for basic amino acids (McDonald and Barrett 1986). There is very little information on the importance and role of these enzymes in meat and meat processing.

Muscle Aminopeptidases. Aminopeptidases appear to be metallo-proteins with very high molecular mass (Table 2.6) and a complex structure. Only five types of separable aminopeptidase activities have been reported: leucyl, arginyl, alanyl, pyroglutamyl and methionyl aminopeptidases. The major aminopeptidases from pork skeletal muscle have been purified and characterized. All of them are active at neutral or basic pH. These enzymes are named on the basis of their preference or requirement for a specific N-terminal amino acid, but their names only indicate the highest rate of hydrolysis on N-terminal amino acid bond. For instance, alanyl aminopeptidase has the highest preference for alanine but is also able to hydrolyze a wide spectrum of amino acids such as aromatic, aliphatic and basic aminoacyl-bonds (Flores *et al.* 1996). This enzyme is the major aminopeptidase in postmortem muscle and accounts for most of the release of free amino acids. Arginyl aminopeptidase hydrolyzes basic amino acids, such as arginine or lysine and this is why it is also named aminopeptidase B (Flores *et al.* 1993). Methionyl aminopeptidase is activated by calcium ions and has a wide spectrum of activity with preference for methionine, alanine, lysine and leucine (Flores *et al.* 2000).

The role of exopeptidases in protein and peptide degradation has received little attention although these enzymes are involved in the latter stages of proteolytic degradation and are responsible for the accumulation of free amino acids during the processing of meat products as will be discussed in Chap. 6.

Muscle Carboxypeptidases. Carboxypeptidases are located in lysosomes, having optimal activity at acid pH. These enzymes generate free amino acids from the carboxy termini of peptides and proteins. Carboxypeptidase A has preference for hydrophobic amino acids while carboxypeptidase B has a wide spectrum of activity (McDonald and Barrett 1986).

Muscle Lipases. The main characteristics of these enzymes are shown in Table 2.7. Lysosomal acid lipase and phospholipase A are located in the lysosomes and constitute the major lipolytic enzymes in muscle. Both enzymes are responsible for the generation of long-chain free fatty acids in postmortem

TABLE 2.6.
MAIN CHARACTERISTICS OF PORK MUSCLE AMINO- AND CARBOXY-PEPTIDASES

Enzyme	EC number	Classification	Location	Molecular mass (KDa)	Optimal pH	Optimal Temp. (°C)
Alanyl aminopeptidase	3.4.11.14.	Cystein/metallo	Cytosol	185	6.5	37
Arginyl aminopeptidase	3.4.11.6.	Cystein/metallo	Cytosol	76	6.5	37
Methionyl aminopeptidase	3.4.11.18.	Cystein	Cytosol	53	7.5	40
Leucyl aminopeptidase	3.4.11.1.	Metallo	Cytosol	324	9.0	45
Pyroglutamyl aminopeptidase	3.4.19.3.	Cystein	Cytosol	24	8.5	37
Carboxypeptidase A	3.4.16.1.	Serin	Lysosome	100	5.2-5.5	37
Carboxypeptidase B	3.4.18.1.	Cystein	Lysosome	50	5.0	37

TABLE 2.7.
MAIN CHARACTERISTICS OF PORK MUSCLE AND ADIPOSE TISSUE LIPOLYTIC ENZYMES

Enzyme	Location	Optimal pH	Optimal Temp. (°C)
Acid lipase	Lysosome	5.0	37
Neutral lipase	Membrane	7.5	45
Phospholipase A	Lysosome	5.0	37
Muscle acid esterase	Lysosome	5.0	30
Muscle neutral esterase	Cytosol	7.5	20
Hormone-sensitive lipase	Adipose tissue	7.0	37
Monoacylglycerol lipase	Adipose tissue	7.0	37
Lipopotein lipase	Adipose tissue	8.5	37
Acid esterase	Adipose tissue	5.0	60
Neutral esterase	Adipose tissue	7.5	45

muscle. Lysosomal acid lipase has a marked preference for the hydrolysis of primary ester bonds of tryacylglycerols at acid pH (4.5–5.5), hydrolyzing di- and monoacylglycerols at a lower rate (Imanaka *et al.* 1985). A deficiency of this enzyme would result in undegraded tryacylglycerols that can accumulate in the muscle (Negre *et al.* 1985). Phospholipase A is also located in lysosomes and regulates the hydrolysis of phospholipids, at positions 1 or 2, at the water-lipid interface. This enzyme is very important in the biochemical pathways involving phospholipids degradation. Acid and neutral esterases have been also identified in the lysosomes and cytosol, respectively (Motilva *et al.* 1992). These enzymes are quite stable and are able to hydrolyze short chain fatty acids from tri-, di- and monoacylglycerols although their activity is restricted due to the short availability of substrate.

Adipose Tissue Lipases. Lipids and carbohydrates are stored as an energy deposit in the adipose tissue and will cover most of the energy needs of the body in times of starvation by mobilizing large amounts (up to 150 g/24 h) of free fatty acids (Belfrage *et al.* 1984). Adipose tissue contains three important lipolytic enzymes: hormone-sensitive lipase, monoacylglycerol lipase and lipoprotein lipase (Toldrá 1992). These enzymes, which exhibit optimal pH in the neutral/basic range, play an important role in the regulation of the overall energy homeostasis in pork.

The hormone-sensitive lipase (HSL) hydrolyzes stored adipocyte lipids prior to their mobilization as free fatty acids (FFA). This enzyme has a high specificity and catalyzes the cleavage of the first ester-bond in triacylglycerols and the resulting diacylglycerols. The 1 (3)-ester bond of acylglycerols is hydrolyzed 4 times faster than the 2-ester bond (Belfrage *et al.* 1984). Tri-, di- and monooleoylglycerol are hydrolyzed at the relative maximal rates 1:10:4. Although the hormone-sensitive lipase is activated by phosphorylation, the triacylglycerol hydrolysis is the rate-controlling step (Fredrikson *et al.* 1981). The monoacylglycerol lipase (MGL) hydrolyzes 1 or 2 monoacylglycerols with no positional specificity. This enzyme is mainly located in the adipocytes, although a little is found in stromal and vascular cells, and is active against medium and long-chain monoacylglycerols (Tornqvist *et al.* 1978). Lipoprotein lipase is an acylglycerol hydrolase, which is located in the capillary endothelium and hydrolyzes the acylglycerol components at the luminal surface of the endothelium (Smith and Pownall 1984). The level of activity of this enzyme is a function of the nutritional and physiological states affecting triglyceride uptake (exercise, pregnancy, etc.). Lipoprotein lipase has a preference for fatty acids at position 1 over those at position 3 (Fielding and Fielding 1980). It hydrolyzes tri-, di- and monoglycerides at the relative maximal rates 2:4:1, showing very low affinity for the monoglyceride with the hydrolysis of few 2-monoglycerides observed. Unsaturated monoacylglycerols are faster hydrolyzed than saturated compounds (Miller *et al.* 1981).

The lipolysis in adipose tissue takes place according to the following reaction (Belfrage *et al.* 1984):

$$\text{Triacylglycerol} \xrightarrow{\text{HSL}} \underset{+\ \text{FFA}}{\text{Diacylglycerol}} \xrightarrow{\text{HSL}} \underset{+\ \text{FFA}}{\text{2-Monoacylglycerol}} \xrightarrow{\text{MGL}} \underset{+\ \text{FFA}}{\text{Glycerol}}$$

This process is controlled by the hormone-sensitive lipase in the first step (rate-limiting) as mentioned above. Monoacylglycerols resulting from this reaction or from the degradation of chylomicron and VLDL-triacylglycerol catalyzed by lipoprotein lipase (Belfrage *et al.* 1984) do not accumulate because of the monoacylglycerol lipase.

Acid and neutral esterases are also present in adipose tissue (Motilva *et al.* 1992) and can participate in the mobilization of stored cholesteryl esters during mobilization of depot lipids and also in the degradation of lipoprotein cholesteryl esters taken up from the plasma (Belfrage *et al.* 1984).

Quality Characteristics

The concept of quality varies in the successive steps of the meat chain, from farm to the table (Fig. 2.3). Pork carcasses of acceptable lean quality and acceptable belly thickness are evaluated for conformation, carcass length and backfat thickness. The carcass yields vary depending on the degree of fatness and degree of muscling. The grade is determined based on both degrees. In the United States, there are four grades. The expected yields are >60.4% for U.S. No. 1, 57.4 to 60.3% for U.S. No. 2, 54.4 to 57.3% for U.S. No. 3 and <54.4% for U.S. No. 4 (Romans *et al.* 1994). The sows are classified into 5 grades: U.S. No. 1, U.S. No. 2, U.S. No. 3, Medium and Cull. Barrow and gilt carcasses will have an acceptable quality of lean and belly thickness. In the case of the European Union, primary and secondary cuts are shown in Fig. 2.4. An example of the respective yields is given in Fig. 2.5.

The variability in meat quality is a major problem within the meat industry. Meat quality is very important for all segments of the industry from producer to consumer. One of the main problems associated with pork quality is the condition known as pale, soft and exudative (PSE), which is characterized by a pale color, soft texture and high drip loss. More recently, a questionable quality known as red, soft and exudative (RSE) has been defined (Warner *et al.* 1997). This meat is difficult to differentiate from normal meat because of its normal color, although it also gives a higher drip loss and soft texture. The other extreme, known as dark, firm and dry (DFD) due to its darker color, firm texture and dry appearance on the external surface, has a lower incidence. However, DFD must be also controlled when processing dry-cured meat products due to its neutral pH, and, if possible, it should be rejected in order to avoid microbial contamination and high water binding.

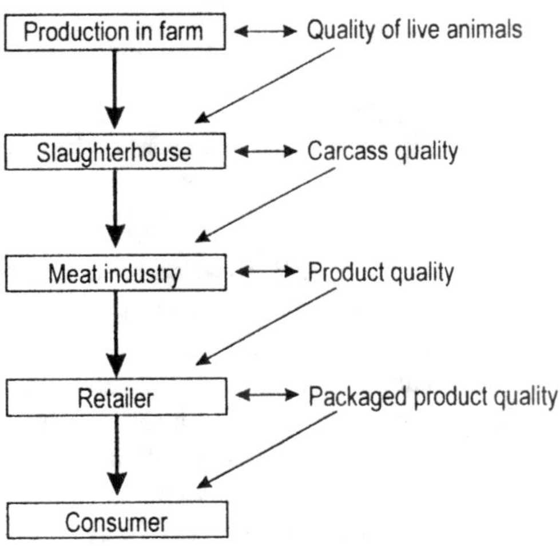

FIG. 2.3. SUCCESSIVE STEPS IN QUALITY APTITUDES THROUGH THE MEAT CHAIN

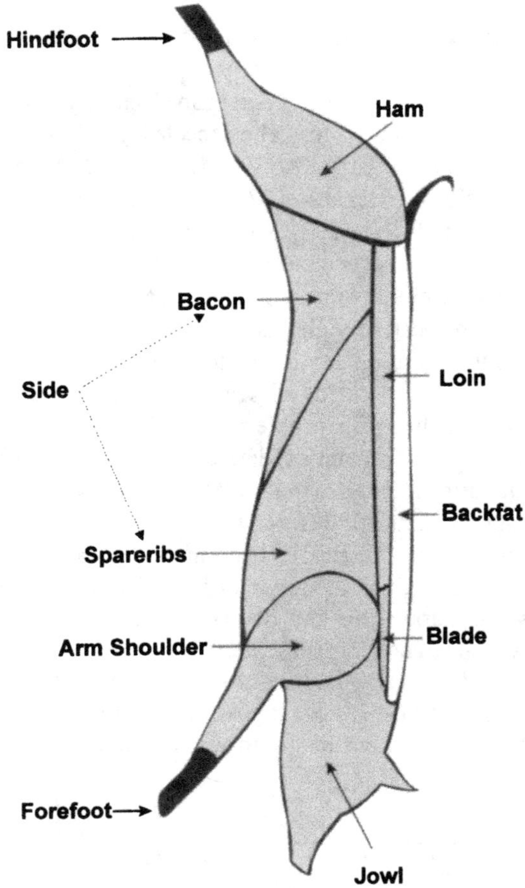

FIG. 2.4. MAIN PRIMARY AND SECONDARY CUTS IN PORK CARCASS
Data may vary depending on type of hog and cutting method.

Both types of meats, PSE and RSE, must be handled very carefully at the factory as they may produce important defects in the final quality of dry-cured meat products, especially in saltiness (due to an excessive salt uptake during the salting stage), poor color (paleness), protein denaturation, an excessive surface dryness and even muscle disjunctions where microorganisms can penetrate and spoil the product. Pork meat classification is usually based on pH, color and drip loss. An example of parameters used for this classification is shown in Table 2.8. The negative effects of exudative meats may be reduced significantly with some preventive measures such as appropriate pretransport handling, lairage, stunning, postmortem temperature and chilling rate.

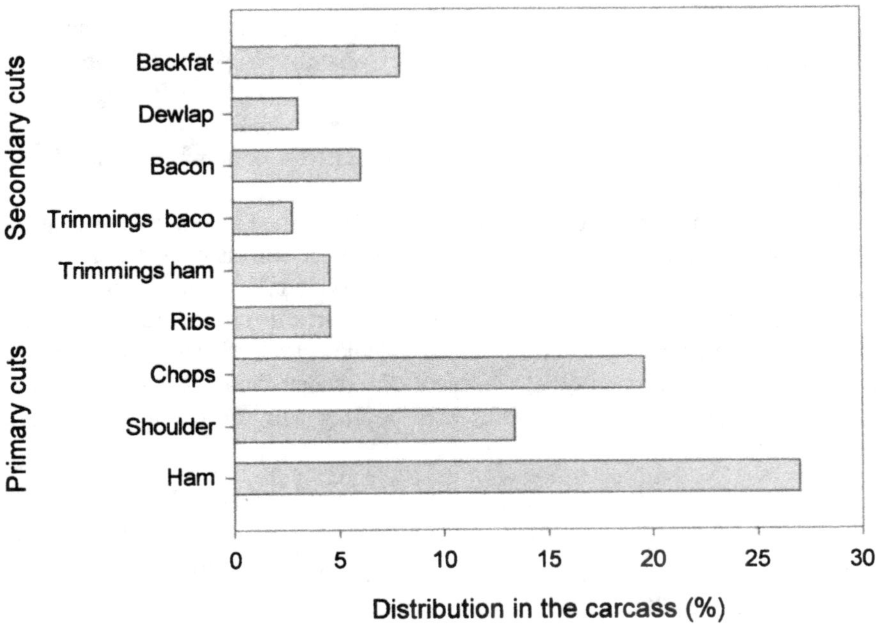

FIG. 2.5. DISTRIBUTION OF THE MAIN PRIMARY AND SECONDARY CUTS IN THE CARCASS (%)
Data may vary depending on type of hog and cutting method.
(Adapted from Armero *et al.* 1999)

TABLE 2.8.
MEAT CLASSIFICATION BASED UPON PH MEASURED AT 45 MIN, 2 H AND 24 H POSTMORTEM IN MUSCLE *SEMIMEMBRANOSUS* (SM), COLOR (L) AND DRIP LOSS

Meat classification	$pH_{45\,min}$ SM	pH_{2h} SM	pH_{24h} SM	L value	Drip Loss (%)
Pale, soft, exudative (PSE)	< 6.0	< 5.8	-	> 50	> 6
Red, soft, exudative (RSE)	< 6.0	< 5.8	-	44-50	> 6
Red, firm, non-exudative (RFN)	≥ 6.0	> 5.8	< 6.0	44-50	< 6
Dark, firm, dry (DFD)	-	-	> 6.0	< 44	< 3

(Adapted from Warner *et al.* 1997 and Flores *et al.* 1999)

The incidence of pork quality problems is still significant. For instance, a 1992 survey of the pork supply in the U.S. revealed that 16% was PSE and 10% was DFD. Only 16% was considered to be of ideal quality (RFN) while the remainder was found to be of somewhat questionable quality (RSE), indicating there has been very little progress in the elimination or minimization of the problem (Cassens 2000).

The quality perceived by consumers may be classified as extrinsic — origin, cost, ethical factors and production systems — or intrinsic — organoleptic characteristics (color, tenderness, flavor, exudation, etc.), nutritional properties (proteins, amino acids, amount of fat, fatty acids profile, cholesterol, vitamins, minerals, trace elements, calories, etc.), hygienic characteristics (low bacterial counts, absence of pathogens, absence of residues and toxins, etc.) and technological aspects (water content, water-holding capacity, connective tissue, marbling, pH, etc.).

REFERENCES

ARMERO, E., NAVARRO, J.L., NADAL, M.I., BASELGA, M. and TOLDRÁ, F. 2001. Lipid composition of pork muscle as affected by sire genetic type. J. Food Biochem. *26*, 91–102.

BELFRAGE, P., FREDRIKSON, G., STRALFORS, P. and TORNQVIST, H. 1984. Adipose tissue lipases. *In* Lipases, B. Borgström and H.L. Brockman (eds.), pp. 365–416. Elsevier, Amsterdam.

BOND, J.S. and BUTLER, P.E. 1987. Intracellular proteases. Ann. Rev. Biochem. *56*, 333–364.

CASSENS, R.G. 2000. Historical perspectives and current aspects of pork meat quality in the USA. Food Chem. *69*, 357–363.

ETHERINGTON, D.J. 1984 The contribution of proteolytic enzymes to postmortem changes in muscle. J. Anim. Sci. *59*, 1644–1650.

ETHERINGTON, D.J. 1987. Conditioning of meat factors influencing protease activity. *In* Accelerated Processing of Meat. A. Romita, C. Valin and A.A. Taylor (eds.), pp. 21–28. Elsevier Applied Sci., London.

FIELDING, C.J. and FIELDING, P.E. 1980. Characteristics of triacylglycerol and partial acylglycerol hydrolysis by human plasma lipoprotein lipase. Biochim. Biophys. Acta *620*, 440–446.

FLORES, M., ARISTOY, M-C. and TOLDRÁ, F. 1993. HPLC purification and characterization of porcine muscle aminopeptidase B. Biochimie *75*, 861–867.

FLORES, M., ARISTOY, M-C. and TOLDRÁ, F. 1996. HPLC purification and characterization of soluble alanyl aminopeptidase from porcine skeletal muscle. J. Agric. Food Chem. *44*, 2578–2583.

FLORES, M., ARMERO, E., ARISTOY, M-C. and TOLDRÁ, F. 1999. Sensory characteristics of cooked pork loin as affected by nucleotide content and post-mortem meat quality. Meat Sci. *51*, 53–59.

FREDRIKSON, G., STRALFORS, P., NILSON, N.O. and BELFRAGE, P. 1981. Hormone-sensitive lipase of rat adipose tissue. Purification and some properties. J. Biol. Chem. *256*, 6311–6320.

GOLL, D.E., OTSUKA, Y., NAGAINIS, P.A., SHANNON, J.D., SATHE, S.K. and MUGURUMA, M. 1983. Role of muscle proteinases in maintenance of muscle integrity and mass. J. Food Biochem. *7*, 137–177.

HERNÁNDEZ, P., NAVARRO, J-L. and TOLDRÁ, F. 1998. Lipid composition and lipolytic enzyme activities in porcine skeletal muscles with different oxidative pattern. Meat Sci. *49*, 1–10.

IMANAKA, T., YAMAGUCHI, M., AHKUMA, S. and TAKANO, T. 1985. Positional specificity of lysosomal acid lipase purified from rabbit liver. J. Biochem. *98*, 927–931.

KOOHMARAIE, M., SEIDEMAN, S.C., SCHOLLMEYER, J.E., DUTSON, T.R. and GROUSE, J.D. 1987. Effect of postmortem storage on Ca^{2+} Dependent proteases, their inhibitor and myofibril fragmentation. Meat Sci. *19*, 187–196.

KOOHMARAIE, M. 1994. Muscle proteinases and meat ageing. Meat Sci. *36*, 93–104.

MATSUKURA, U., OKITANI, A., NISHIMURO, T. and KATO, H. 1981. Mode of degradation of myofibrillar proteins by an endogenous protease, cathepsin L. Biochim. Biophys. Acta *662*, 41–47.

MCDONALD, J.K. and BARRETT, A.J. (eds.) 1986. Mammalian Proteases. A Glossary and Bibliography. Vol. 2. Exopeptidases. Academic Press, London.

MILLER, C.H., PARCE, J.W., SISSON, P. and WAITE, M. 1981. Specificity of lipoprotein lipase and hepatic lipase toward monoacylglycerols varying in the acyl composition. Biochim. Biophys. Acta *665*, 385–392.

MOTILVA, M-J., TOLDRÁ, F. and FLORES, J. 1992. Assay of lipase and esterase activities in fresh pork meat and dry-cured ham. Z. Lebensm. Unters Forsch. *195*, 446–450.

NEGRE, A.E., SALVAYRE, R.S., DAGAN, A. and GATT, S. 1985. New fluorimetric assay of lysosomal acid lipase and its application to the diagnosis of Wolman and cholesteryl ester storage diseases. Clin. Chim. Acta *149*, 81–88.

OKITANI, A., NISHIMURA, T. and KATO, H. 1981. Characterization of hydrolase H, a new muscle protease possesing aminoendopeptidase activity. Eur. J. Biochem. *115*, 269–274.

OUALI, A. and TALMANT, A. 1990. Calpains and calpastatin distribution in bovine, porcine and ovine skeletal muscles. Meat Sci. *28*, 331–348.

RICO, E., TOLDRÁ, F.and FLORES, J. 1990. Activity of cathepsin D as affected by curing agents, Aw and pressure. Z. Lebensm. Unter. Forsch. *191*, 20–23.

RICO, E., TOLDRÁ, F. and FLORES, J. 1991. Effect of dry-curing process parameters on pork muscle cathepsins B, H and L activities. Z.Lebensm. Unter. Forsch. *193*, 541–544.

ROBSON, R.M., HUFF-LONERGAN, E., PARRISH JR, F.C., HO, C-Y. STROMER, M.H., HUIATT, T.W., BELLIN, R.M. and SERNETT, S.W. 1997. Postmortem changes in the myofibrillar and other cytoskeletal proteins in muscle. Proceedings of the 50th Annual Reciprocal Conference. Vol. 50, pp. 43–52. Ames, Iowa.

ROMANS, J.R., COSTELLO, W.J., CARLSON, C.W., GREASER, M.L. and JONES, K.W. 1994. The Meat We Eat. Interstate Publishers Inc., Danville, IL.

SCHWARTZ, W.N. and BIRD, J.W.C. 1977. Degradation of myofibrillar proteins by cathepsins B and D. Biochem. J. *167*, 811–820.

SENTANDREU, M.A. and TOLDRÁ, F. 1998. Biochemical properties of dipeptidylpeptidase III purified from porcine skeletal muscle. J. Agric. Food Chem. *46*, 3977–3984.

SENTANDREU, M.A. and TOLDRÁ, F. 2000. Purification and biochemical properties of dipeptidylpeptidase I from porcine skeletal muscle. J. Agric. Food Chem. *48*, 5014–5022.

SENTANDREU, M.A. and TOLDRÁ, F. 2001a. Importance of dipeptidyl peptidase II in postmortem pork muscle. Meat Sci. *57*, 93–103.

SENTANDREU, M.A. and TOLDRÁ, F. 2001b. Dipeptidyl peptidase IV from porcine skeletal muscle: purification and bochemical properties. Food Chem. *75*, 159–168.

SMITH, L.C. and POWNALL, H.J. 1984. Lipoprotein lipase. *In* Lipases, B. Borgström and H.L. Brockman (eds.), 263–305. Elsevier, Amsterdam, The Netherlands.

TOLDRÁ, F. 1992. The enzymology of dry-curing of meat products. *In* New Technologies for Meat and Meat Products, F.J.M. Smulders, F. Toldrá, J. Flores and M. Prieto (eds.), pp. 209-231. Audet, Nijmegen, The Netherlands.

TORNQVIST, H., NILSSON-EHLE, P. and BELFRAGE, P. 1978. Enzymes catalyzing the hydrolysis of long-chain monoacylglycerols in rat adipose tissue. Biochim. Biophys. Acta *530*, 474–486.

WARNER, R.D., KAUFFMAN, R.G. and RUSSELL, R.L. 1997. Muscle protein changes post mortem in relation to pork quality traits. Meat Sci. *33*, 359–372.

ZEECE, M.G. and KATOH, K. 1989. Cathepsin D and its effects on myofibrillar proteins: A review. J. Food Biochem. *13*, 157–178.

CHAPTER 3

MANUFACTURING OF DRY-CURED HAM

Dry-cured ham originated as a meat preservation process for times of scarcity. However, the expanded use of refrigeration has reduced this need although the product has increased in consumer acceptance. In this sense, the process has experienced different modifications and improvements in order to obtain a flavorful and attractive meat product. However, the processing technology is largely empirical, following traditional know-how transmitted by manufacturers from generation to generation. The assessment of quality is primarily subjective and depends on experienced judges.

For many years, the lack of scientific knowledge of how chemical and biochemical mechanisms related with flavor and texture development has slowed the technological evolution in relation to other food processes. Numerous biochemical reactions, mainly affecting proteins and lipids, take place during the dry-curing process, especially along the ripening period contributing to the development of an adequate texture and a characteristic flavor. A significant increase in the knowledge of these reactions has been obtained in the last decade (Parolari 1996; Toldrá and Flores 1998; Toldrá 1998).

There are many factors affecting the quality of dry-cured hams. The raw materials and the ripening conditions have a special influence on the final texture and flavor (as discussed in Chap. 10) and on its homogeneity through the whole piece. The most important muscles in a ham are shown in two cross sections in Fig. 3.1. There are important technological variations that result in a large variability in the final quality. Some of the most important are: the genetic types of pigs (autochthonous or modern crossbreeds), the age at slaughter (5–18 months), the type of feeding (composition, extensive or intensive) and the processing technology (type of salting, post-salting duration, processing conditions, etc.).

There are many types of dry-cured hams in the Mediterranean area. The characteristics of each ham, as mentioned above, depends on the particular pig breed, age and feed as well as the type of process. Some of the most important dry-cured hams are Spanish Iberian and Serrano hams, Italian Parma and San Daniele prosciuttos and French Bayonne ham. These hams are usually consumed raw with no further smoking or cooking. Other smoked dry-cured hams are cooked before consumption and are produced in other areas such as the country-style ham in the U.S. and the Westphalia ham in Germany.

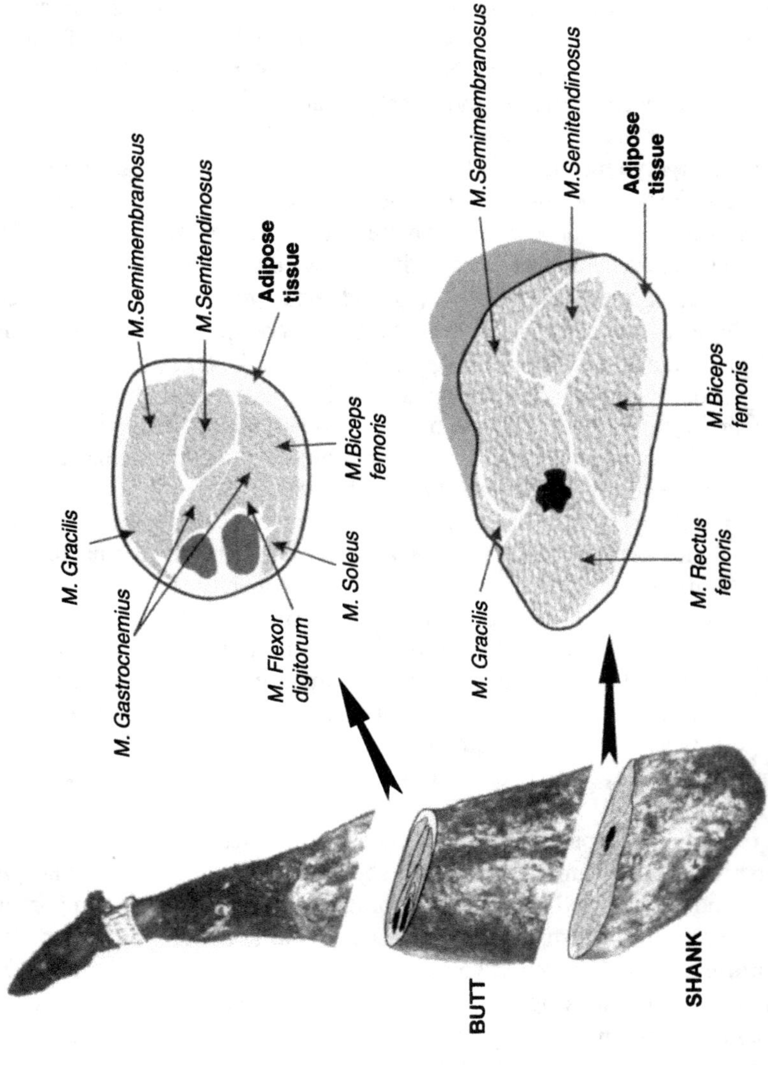

FIG. 3.1. CROSS-SECTION OF DRY-CURED HAM AT BUTT AND SHANK LOCATIONS SHOWING THE MOST RELEVANT MUSCLES

Ingredients and Additives

Meat. The hams are usually classified by pH, weight and sometimes cover fat thickness before entering the salting stage. It is necessary to control the hygiene, with periodical bacteriological controls, and check the temperature and pH of hams at reception. If possible, it would be convenient to measure the pH at 1 h postmortem to discern exudative (PSE and RSE) hams from normal hams. DFD hams, which can be screened with pH measurements at 24 h postmortem (pH_{24h}), should be rejected in order to avoid microbial contamination and high water binding due to its neutral pH. The migration of salt is faster in these hams. In this sense, hams with pH at 1 h postmortem (pH_{1h}) above 6.0, or pH_{24h} between 5.6 and 6.1 should be chosen. The exudative hams, which have a lower water binding capacity, may reach weight losses around 4% higher than normal hams (Maggi and Oddi 1988). They present a higher amount of moisture on the surface that helps the dissolution of the added solid salt and facilitates its penetration into the ham. So, PSE/RSE meats may present important defects in the final quality such as serious problems in saltiness (because of an excess of salt uptake during the salting stage), color (paleness) and protein denaturation. Exudative hams may also present an excessive surface dryness and even muscle disjunctions where microorganisms can penetrate and spoil the product. However, some PSE hams may be processed under controlled conditions without significantly affecting the final quality.

Today, most of the standard quality hams are produced from young light pigs, around 110–120 Kg live weight. However, high quality hams are produced from older heavy pigs with well-known specific crossbreeds. The meat from older pigs, which contains a higher amount of myoglobin and a different enzyme profile than younger pigs, is preferred because of a more intense cured color, a higher fat content and a better flavor profile. However, from a commercial point of view, it is not always possible to use older pigs because of their higher price and their lower added value of the rest of the meat.

Fat. It is recommended to use fresh hams or those not stored for a long time because of the freshness of fat tissue. Endogenous lipases, which are active at low temperatures, may act and generate free fatty acids (lipolysis) during cold storage and even during frozen storage. The amount of polyunsaturated fatty acids is also important because they are prone to oxidation and can develop off-flavors. The composition in fatty acids, which is of primary importance for flavor development, depends on the feed (as will be discussed in Chap. 10). This fact can originate strong discussions among farmers and ham producers. Periodical controls of quality, like iodine index (as an indicator of unsaturation) and acid index (as an indicator of freshness), are recommended. Soft fats, which usually contain a high percentage of polyunsaturated fatty acids, must be handled with extreme care in order to avoid an excess of oxidation and off-flavor development.

Curing Agents. Salt, nitrate and nitrite are the main curing agents. Salt is always present in cured meat products and has several roles in the final quality: preservation by bacteriostatic effect, inhibiting the growth of undesirable microorganisms; contribution to its characteristic salty taste; and increase in myofibrillar protein solubility. The usual final amounts of salt found inside the ham are around 4–6% although higher values such as 8–9% may be found in certain hams.

The addition of nitrate, previously formulated in a curing salt or combined with nitrite, is a common practice. Nitrate can be added as sodium or potassium salt and is generally used for a long term product. Nitrite is produced from nitrate by the natural microbial flora present in the ham (Micrococcaceae and others) through the nitrate reductase activity, or, in minor cases, added sodium or potassium salt are part of the curing salt. The amounts of nitrate and/or nitrite are restricted to those necessary for protection against botulism (Cassens 1995). The European Directive 95/2/CE (1995) allows maximum and residual amounts of nitrates and nitrites in meat products. The maximum allowed amount is 150 ppm (if alone) or 300 ppm when combined (nitrate+nitrite), and the residual values should be below 50 ppm (if alone) or 250 ppm (if combined). The maximum amount allowed in the U.S. for sodium nitrite is 156 ppm (1/4 ounce per 100 pounds of meat).

The reduction of nitrate to nitrite is slow due to the low bacterial counts but allows the diffusion of nitrate through the whole piece. The reduction of nitrite to nitric oxide is favored at slight acid medium (pH 5.6–6.0) with the presence of reducing substances such as ascorbic or erythorbic acids usually added as curing adjuncts. Nitric oxide is very reactive and interacts with protein and other meat components. It reacts with myoglobin forming nitrosomyoglobin, an essential reaction for the development of the characteristic bright red cured color. About 10 to 40% of total myoglobin is transformed into nitrosomyoglobin (Frentz and Zert 1990). The main reactions are the following:

KNO_3 → KNO_2 catalyzed by nitrate reductase
HNO_2 → HNO2 in slight acid medium
HNO_2 → NO catalyzed by reducing agents in slight acid medium
NO + MYOGLOBIN → NITROSOMYOGLOBIN

The pink pigment is relatively stable due to the presence of salt and slight acid media, but the color may change by microbial action, oxidative agents (i.e., peroxides formed by bacteria) or light. It is important to keep strict hygienic conditions to avoid microbial contamination and protect the product from light.

Nitrite also acts as a preservative inhibiting the growth of undesirable microorganisms, with specific protection against *Clostridium botulinum*, and also contributes to the typical cured flavor by regulating the oxidative processes (Cassens 1997). The nitrate and nitrite residues are generally low in the finished product.

Ascorbic and erythorbic acids, or their sodium salts, are added as reducing agents to speed up the curing reactions. They facilitate the formation of nitric oxide, exert antioxidant activity stabilizing color and flavor and inhibit the formation of nitrosamines. Sugars, like glucose or saccharose, may be added for a slight development of the natural microbial flora and for a milder salty taste.

Traditional Processing

The traditional processing of dry-cured ham was transmitted from generation to generation and involved several stages: salting, post-salting, and drying/aging, the most important. By the end of the process, hams were stabilized by the presence of salt and reduction in water activity and had developed typical sensory characteristics. Ancient practices involved rearing pigs at home and slaughtering them around the end of autumn or early winter, just at the beginning of the cold weather. Hams were either individually rubbed with dry salt or left to stand in piles — layer by layer, up to six layers — and then fully surrounded by salt (Fig. 3.2.). This method of salting is still used in some modern processing facilities, mainly in Spain, San Daniele (Italy) and southwest of France. Salting and post-salting extended for a period of time along with the coldest winter months (December to February) while ripening/drying took place during the spring and summer, respectively. The product was then ready for consumption. The production sites were usually located in the middle-high mountains where the climate is cool and dry and favors a correct drying. The air circulation in the drying room, and subsequently the drying rate, was manually regulated by controlling the opening/closure of the windows, depending on the weather conditions. The operator decided, on a daily basis and according to the weather conditions, the necessary aeration by checking the moisture content of the hams by visual/tactile assessment.

Modern Processing Technology

Today, most of the factories producing hams are well equipped with powerful refrigeration units and chambers with computer-controlled temperature, air speed and relative humidity. Computers are able to reproduce the four seasons of the year in controlled drying chambers. The stages in processing dry-cured ham are shown in Fig. 3.3. The process may be fast or slow, depending on the desired final quality. In general, milder processes over longer periods of

FIG. 3.2. TRADITIONAL SALTING OF HAMS
(Courtesy of Jamones Segorbe, S.L., Segorbe (Castellón), Spain)

time produce a better quality ham than when hams are processed within a few months. The longer the ripening period is, the more time is left for the endogenous enzyme action and the accumulation of flavor compounds (Toldrá 1998). A flow diagram of the processing is also shown in Fig. 3.4. The process may vary depending on the type of ham, local traditions, climate, etc., as reflected in Tables 3.1 and 3.2. A basic outline of the procedures follows.

Reception. Pork legs and shoulders arrive at the factory either fresh or frozen with the bone in. They must meet specifications of weight, pH and fat thickness and composition, which are all controlled at the entry of the factory by trained experts. They may be grouped by weight and pH to facilitate the control of salting. Each region has its own specifications. For instance, fat thickness must be equal or higher than 10 mm, the temperature inside hams must be below 5C and contents in linoleic acid below 15% in Bayonne, France. The hams are trimmed to make them round, and part of the skin is removed in different types of hams depending on the typical traditions (i.e., it is cut as a V form in Spain). Hams are given a lot and batch number to facilitate the traceability of the product through the whole process. Hams are placed in plastic or stainless steel shelves and held for one day under refrigeration (2–4C) in order to get a uniform temperature. Then, hams are placed on a steel belt with pressing rollers for bleeding. An adequate elimination of remaining blood from veins and arteries is essential to avoid product losses at the end of the process.

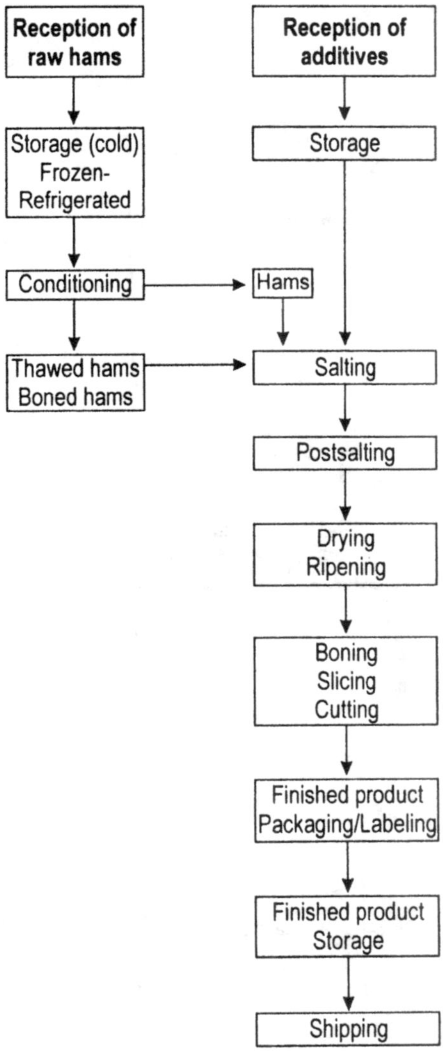

FIG. 3.3. MAJOR STEPS IN THE MANUFACTURE OF A TYPICAL DRY-CURED HAM

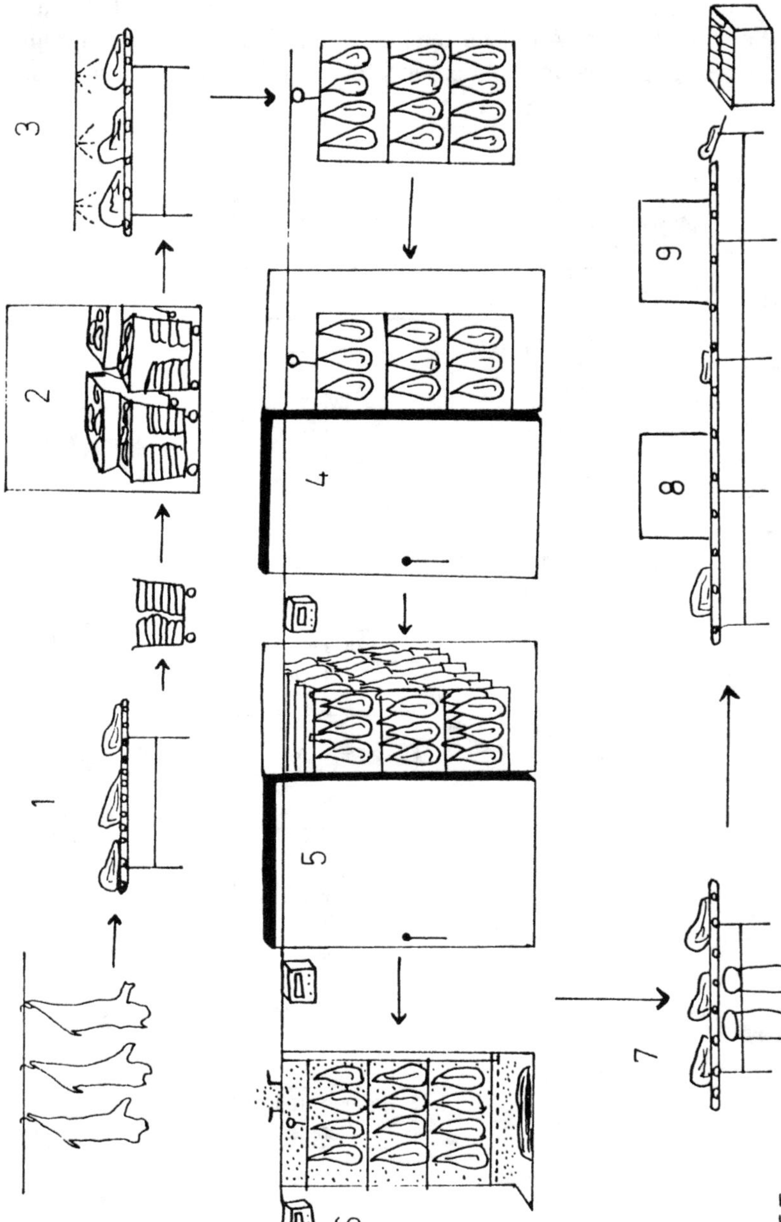

FIG. 3.4. FLOW DIAGRAM IN THE PROCESSING OF DRY-CURED HAMS. (1) RECEPTION, (2) SALTING, (3) RINSING, (4) POSTSALTING, (5) DRYING, (6) SMOKING (OPTIONAL), (7) DEBONING, (8) PRESSING, (9) SLICING AND PACKAGING

TABLE 3.1.
EXAMPLE OF SALTING AND POST-SALTING CONDITIONS IN THE PROCESSING OF SERRANO, IBERIAN, PARMA, FRENCH, BAYONNE AND COUNTRY-STYLE DRY-CURED HAMS

Stage		Serrano[1]	Iberian[2]	Parma[3]	French[4]	Bayonne[5]	Country-style[6]
Skin removal		Yes	Yes	No	No	No	No
Salting	***1st phase***						
	T (°C)	0-4	0-4	1-4	1-3	1-4	2-4
	t (days)	10-12	10-12	5-6	14-21	7	40-50
	RH (%)	75-95	75-95	75-90	85-95	85-95	80-95
	2nd phase						
	T (°C)	—	—	1-4	—	2-4	—
	t (days)	—	—	21	—	14	—
	RH (%)	—	—	70-80	—	80-90	—
Post-salting	***1st phase***						
	T (°C)	4-6	4-6	1-4	1-4	4	10-12
	t (days)	40-60	40-60	14	14-21	60-65	15
	RH (%)	70-95	70-95	50-60	70-85	65-85	75
	2nd phase						
	T (°C)	—	—	1-4	4		—
	t (days)	—	—	70	21-35		—
	RH (%)	—	—	70-90	75-80		—
Smoking	***T (°C)***	No	No	No	No	No	38
	t (hours)						24

(Adapted from [1]Toldrá *et al.* 1997, [2]Toldrá and Flores 1998, [3]Parolari 1996, [4]Frentz and Zert 1990, [5]Monin *et al.* 1997, [6]Marriott *et al.* 1992)

TABLE 3.2.
EXAMPLE OF DRYING-AGING CONDITIONS IN THE PROCESSING OF SERRANO, IBERIAN, PARMA, FRENCH, BAYONNE AND COUNTRY-STYLE DRY-CURED HAMS

Stages		Serrano[1]	Iberian[2]	Parma[3]	French[4]	Bayonne[5]	Country-style[6]
Ripening-Drying	***1st phase***						
	T (°C)	6-16	6-16	15-18	20-25	20-22	25-30
	t (days)	>45	>90	180-330	2-4	4	30-90
	RH (%)	70-95	60-80	65-75	75-85	75-85	65
	2nd phase						
	T (°C)	16-24	16-26	—	14-15	12-15	—
	t (days)	>35	>90	—	40-55	180	—
	RH (%)	70-95	55-85	—	75-80	75-80	—
	3rd phase						
	T (°C)	24-34	12-22	—	12-16	—	—
	t (days)	>30	>115	—	>90	—	—
	RH (%)	70-95	60-90	—	85	—	—
	4th phase						
	T (°C)	12-20	—	—	—	—	—
	t (days)	>35	—	—	—	—	—
	RH (%)	70-95	—	—	—	—	—
TOTAL TIME	***(days)***	>215	>365	300-455	157-298	>270	85-145

(Adapted from [1]Toldrá *et al.* 1997, [2]Toldrá and Flores 1998, [3]Parolari 1996, [4]Frentz and Zert 1990, [5]Monin *et al.* 1997, [6]Marriott *et al.* 1992)

If frozen hams are used, they must be left to thaw until internal temperatures above -5 to -3C before entering the process. These frozen/thawed hams, which are also grouped by weight, will be subjected to different salting conditions.

Salting. The Spanish Serrano and Iberian hams are pre-salted for the addition of nitrate. Hams are nitrified either manually or in rotary drums with a mixture of curing ingredients, mainly salt and nitrate, in the form of a curing salt (sodium chloride with 4% potassium nitrate). In some cases, like French and country-style hams, the curing salt, which already contains salt and potassium nitrate or sodium nitrite, is used during the salting stage. In Bayonne hams, curing salt (salt + 0.1% sodium nitrite) can be used to a maximum of 20 g/Kg ham. Nitrate can be used to a maximum amount of 2 g/kg ham if alone or 1 g/kg ham if curing salt is also used. In other cases, like Parma hams (the use of nitrate was banned in 1993), there is no addition of nitrate, and this step is thus unnecessary.

There are two main procedures for the salting stage, depending on the control of the salt added to the hams: undetermined salt supply and exact salt supply.

Undetermined Salt Supply. This procedure is mainly used in Spain and part of France and Italy. The hams are completely surrounded by salt (rough sea salt or refined mineral salt) and placed layer by layer into stainless steel bins with holes at the bottom for dripping removal (Fig. 3.5). The hams are left fat side down to allow the curing ingredients to diffuse under the conditions. Alternatively, sea salt or mineral salt (coarse to medium grain) may be rubbed onto the lean surface, and the hams are then placed on shelves. The outer surface is very important because penetration of salt takes place only on the lean meat area. The usual conditions for salting are 1.1 day per kg of ham at 3–4C and high relative humidity to prevent excessive dehydration of the surface. A second salting, which may be necessary for heavy hams, is a common practice in several types of hams. For instance, French hams are washed and salted again for an additional period of time. Weight losses may reach 3–4% at the end of this stage (Fig. 3.6) for a typical evolution during the processing of different types of hams. When there is an abundant surface moisture, especially in exudative hams and frozen/thawed hams, which can facilitate the dissolution of salt and its penetration into the meat, the time is reduced by 2 days.

Exact Salt Supply. The exact amount of salt per kg of ham is added on the lean surface and hand-rubbed. This process is longer because it is necessary to wait until all the salt has been absorbed. It may take between 14 and 21 days, depending on the size of the ham. The process is complicated and the hygrometry must be well controlled. In the case of Parma hams, the lean surface is rubbed with medium-grain salt (20–30 g per kg of ham) and the skin is rubbed with 10–20 g of wet salt (coarse salt with 20% of water) per kg of ham (Parolari 1996). The salt is renewed at the beginning of the second phase, and the relative humidity is decreased to favor mild dehydration of the surface moisture.

FIG. 3.5. MODERN SALTING OF HAMS IN VATS

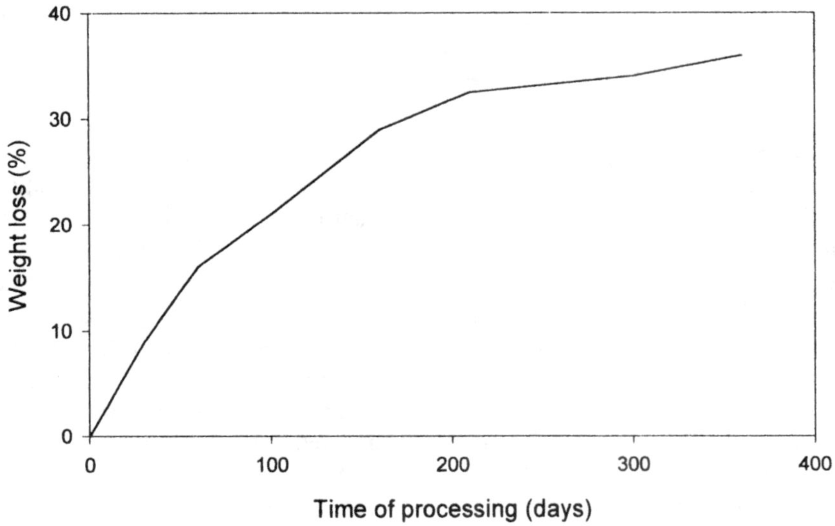

FIG. 3.6. EXAMPLE OF EVOLUTION OF WEIGHT LOSS (IN %) IN RELATION TO THE INITIAL FRESH WEIGHT ALONG THE PROCESSING OF SPANISH SERRANO DRY-CURED HAM
(Toldrá 1991, Unpublished data)

Post-salting or Resting. The hams are washed with water by rinsing and brushing for the removal of residual salt from the external surface. Hams are then left on platforms or hung on drying rails or steel shelves and placed in chambers for salt equalization. The complete homogeneous distribution of the curing agents through the entire piece usually takes 1–2 months, depending on the ham size, ratio of lean surface to mass, pH, presence of intramuscular fat (which constitutes a barrier to salt diffusion) and temperature of the chamber (which regulates the diffusion rate and type of process). The relative humidity progressively decreases as the resting stage progresses. The weight losses are around 4–6%. French hams are typically heated at 22–24C for a week in order to dry hams after washing and fix the color.

Smoking. Certain types of hams, such as the American country-style or the German Westphalia ham, are heated/smoked for a few hours after the resting period. Smoking is also one of the oldest preservation processes. It can also be applied, sprayed or atomized on the surface of the ham. The application of smoking depends on the type of ham and the country. Smoking is more common in places where the application of drying is more difficult, like northern countries. Smoking gives a characteristic flavor to the ham and protects its surface from mold or yeast growth due to the bactericide and bacteriostatic effect of smoke compounds. On the other hand, formaldehyde and other possible carcinogenic substances contained in the smoking process may have adverse effects on health (Bem *et al.* 1995).

Ripening/Drying. In order to develop the characteristic dry-cured flavor and texture, the hams in the racks are moved to the drying chambers where they are dried and ripened at different time-temperature conditions. Traditionally, the ripening/drying was operated in natural rooms located in places with dry weather conditions, such as in the mountains of Spain, in the dry winds from the Alps or Pyrenees or in the valleys of Parma or Bayonne. Today, most of the ripening/drying is performed in well-equipped chambers with full computer control of temperature, air speed and relative humidity. Hams must not touch each other so air can circulate (Fig. 3.7). Aeration must be uniform to ensure that temperature and relative humidity are homogeneous through the entire room but not too fast (less than 0.02–0.03 m/s), as the ham must not dry too quickly. The excessive dryness of the ham surface retards the transport of water to the outer surface and thus evaporation. It may also cause disjunctions among muscles where microorganisms might penetrate into the ham. Aeration may be performed either by convection, where the warm air introduced at the bottom of the chamber is lighter than cold air and produces a slow movement of the air, or by mechanical ventilation where fans and turbulence are created by baffles on walls.

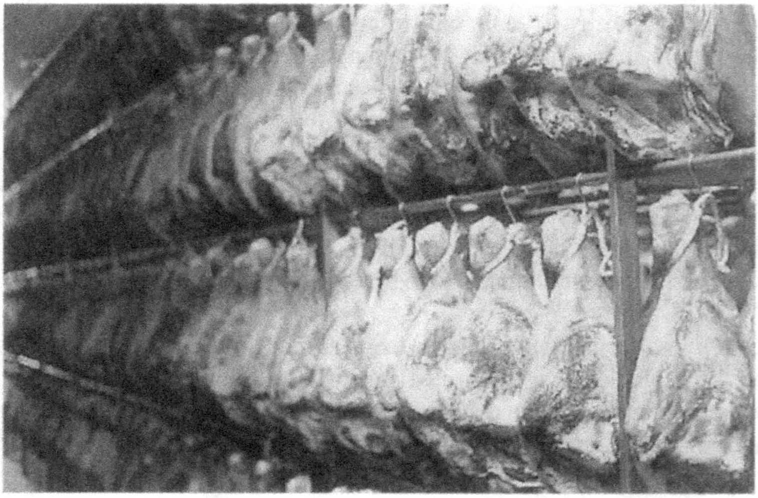

FIG. 3.7. CONTROLLED TEMPERATURE AND HUMIDITY CHAMBER FOR THE DRY-CURING OF HAMS
(Courtesy of Jamones Segorbe, S.L., Segorbe (Castellón), Spain)

Mild temperatures over an extended period of time allow the action of the muscle enzymes and the generation of desirable taste and aroma compounds. The moisture content of the hams is appreciated, not only by visual and tactile assessment, but also by weighing. Spanish hams experience a progressive increase in temperature along the ripening stage. French hams are usually heated to 22–26C for a few days just after the resting period as mentioned above. The enzymatic reactions are accelerated, and the drying is performed in two phases. Once the expected moisture loss is achieved, usually around 7–9 months of processing, the hams are manually smeared with a layer of lard in order to prevent excessive dehydration. Lard also avoids mold and yeast contamination on the outer surface of the ham. The weight loss of this stage is around 20–25%, and the total weight loss in relation to the initial raw ham weight may reach 32–36% at the end of the process (Fig. 3.6).

Drying results in a harder texture of the ham. The drying rate is affected by several factors as summarized in Fig. 3.8. Some of them are intrinsic to the ham such as pH (low pH favors water loss), amount of intramuscular fat (to constitute a barrier to water diffusion), weight of the ham (which extends the time necessary to achieve the desired water loss percentage) and water content in the ham. The outer surface is very important because evaporation takes place only on the lean meat area. Therefore, the ratio of ham surface to mass is important, but so is the presence of molds that can grow and impede evaporation by closing the pores on the surface. The chamber conditions (temperature, air speed and relative humidity) for correct water evaporation are essential for a successful drying.

Final Product. The sale of the entire piece of ham, including the bones, makes individual control by the probe and sniff technique necessary. This technique involves the insertion of a small probe (a thin horse bone 20 cm length × 2–3 mm diameter) known as horse fibula. This fibula is inserted in specific areas of the ham that are prone to spoilage, such as the tibia-femur joint in deep shank, and is immediately sniffed by an expert panelist to detect any off-odor. The percentage of product rejections have substantially decreased over the last few years thanks to the expanded use of adequate refrigeration equipment in the factories and the correct salting and reduction in water activity. The commercial availability of electronic noses, as well as probes connected to these instruments in recent years, will facilitate a rigorous and objective control in the immediate future. These instruments must be calibrated according to each factory's specifications, and, afterwards, they can classify hams correctly. Preliminary research on the classification of two lots of dry-cured hams with different quality (long and short processing) showed promising results (Spanier *et al.* 1999). There is no doubt that the use of these instruments for the quality control of hams will expand in the next few years.

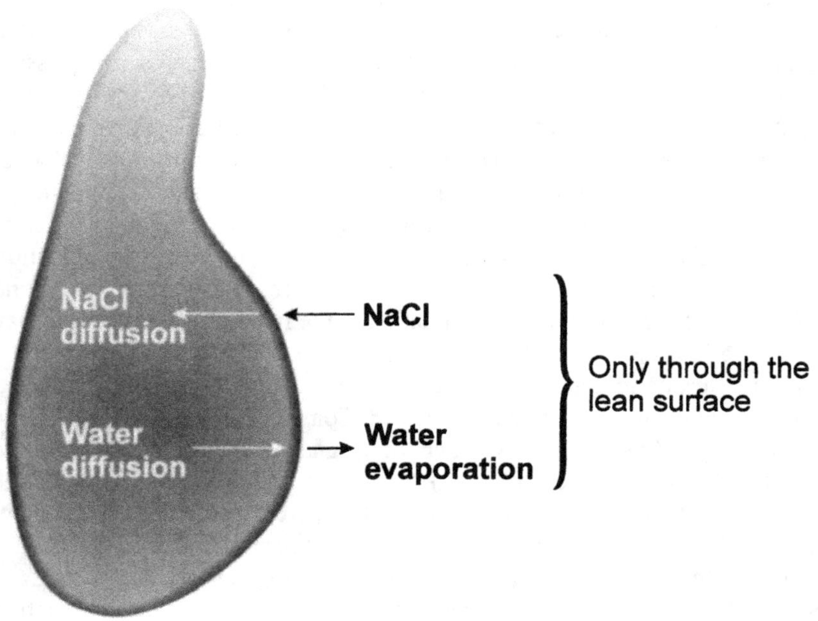

Main factors affecting drying rate

Inner ham:	pH
	Amount of intramuscular fat
	Weight of ham
	Water content in the ham
Outer ham:	Ratio of ham surface to mass
	Presence of molds on surface
Chamber environment:	Temperature
	Relative Humidity
	Air rate

FIG. 3.8. SCHEME OF DRYING AND MAIN FACTORS AFFECTING DRYING RATE

Once the hams pass the sniff test, they are placed in cellars for final maturation during a few weeks or months, depending on the type of ham. The highest quality hams are usually sold whole, including the bone and foot, and cut into thin slices (Fig. 3.9). The rest of the hams are boned and stored for sale in several ways. Bones are removed by expert operators, and hams are compressed to achieve a certain kind of ham shape. Boned hams may be cut into pieces or sliced and stored either in vacuum-packages or under controlled-atmosphere. The whole boned ham may be distributed for slicing in retail shops. Sales of pre-sliced, packaged dry-cured ham have quickly increased in recent years, reaching up to 50% of the total commercial distribution.

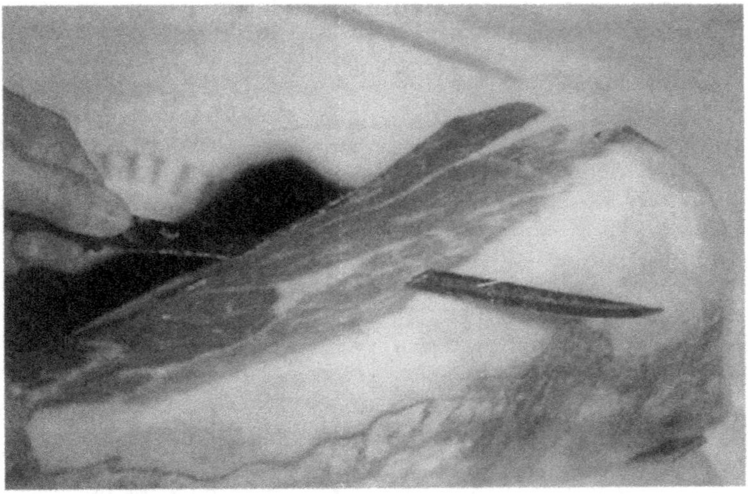

FIG. 3.9. DETAIL OF THE CUT INTO THIN SLICES AND APPEARANCE OF DRY-CURED HAM SHOWING AN INTENSE MARBLING

Changes During Dry-Curing

The main physical and chemical changes taking place during the processing of dry-cured ham are schematized in Fig. 3.10 and described below.

Moisture Content and Water Activity

The kinetics of chemical, enzymatic and microbial reactions are dependent on moisture content and water activity. An example of the evolution of moisture content in the inner and outer muscles of the ham along the processing is shown in Fig. 3.11. There is a gradient between inner and outer muscles with a trend to equilibrate toward the end of the process. The evolution of the moisture content in the muscle *Biceps femoris* during the processing of different dry-cured hams is shown in Fig. 3.12 (A) and the evolution of a_w in Fig. 3.12 (B). In all cases, the highest water loss takes place during the final drying/ripening stage. However, the water diffusion/evaporation is a slow and difficult process. When there is a concentration gradient in a media, a mass transfer is produced and the diffusion follows the Fick's law according to the following formula:

$$(X - X_0) = D\,(C - C_0)$$

where X is distance (m), C the moisture content (Kg/m^3) and D is the diffusivity coefficient (m^2/s). This coefficient cannot be considered as constant because there is a high variability due to the different conditions, intrinsic characteristics

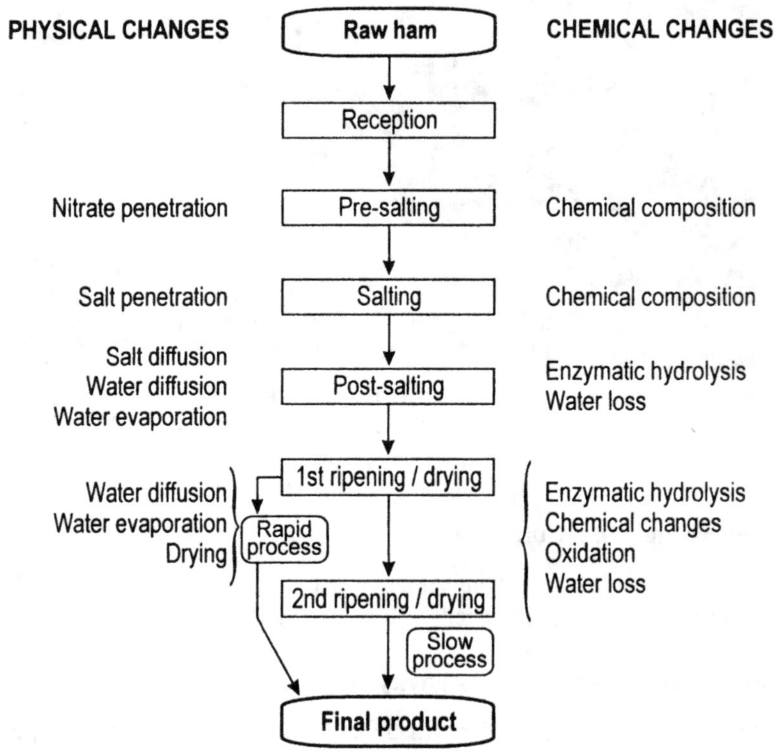

FIG. 3.10. MAJOR STEPS IN THE MANUFACTURE OF A TYPICAL DRY-CURED HAM WITH EMPHASIS ON MAIN PHYSICAL AND CHEMICAL CHANGES

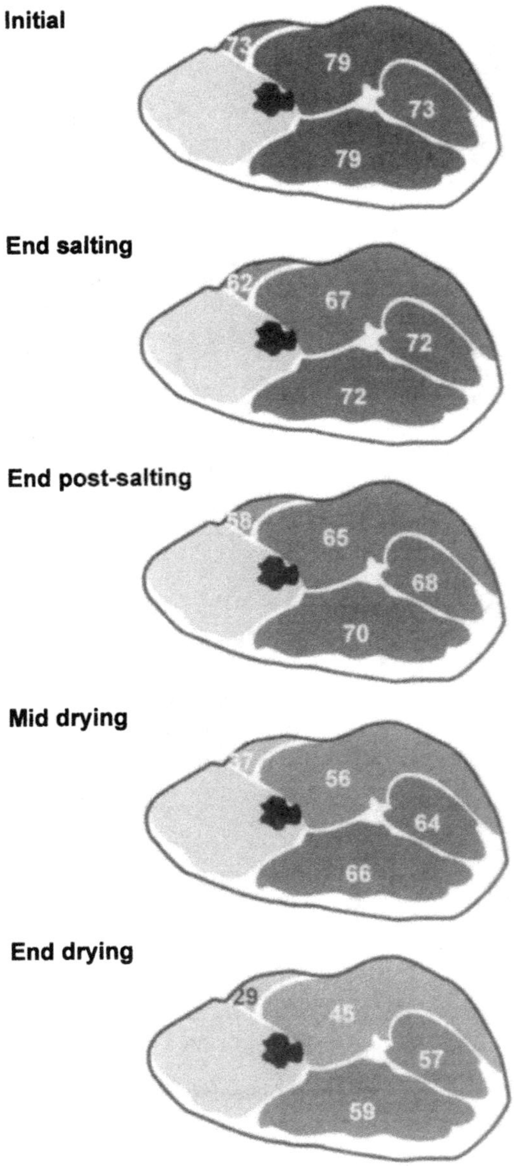

FIG. 3.11. EXAMPLE OF THE EVOLUTION OF MOISTURE IN DIFFERENT INNER AND OUTER MUSCLES DURING THE PROCESSING OF DRY-CURED HAM
Moisture content is expressed as a percentage given in number
(Adapted from Arnau *et al.* 1995 and Toldrá 1991, unpublished data)

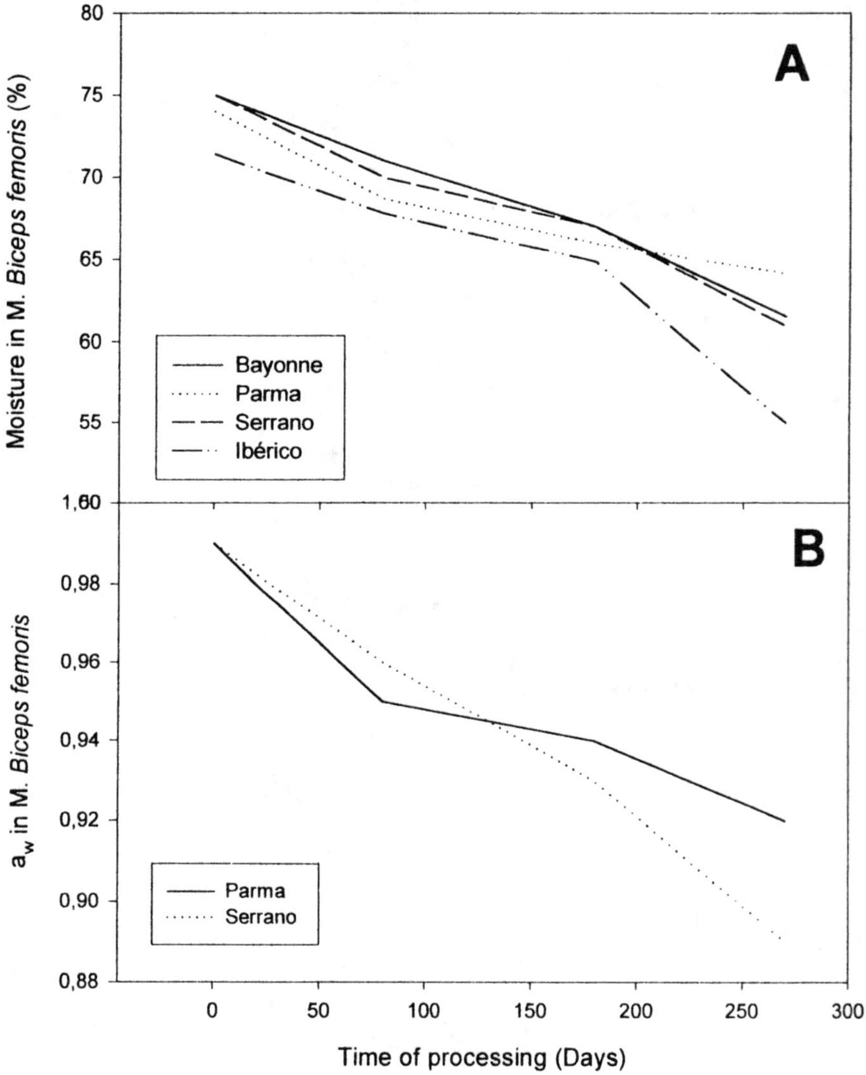

FIG. 3.12. EVOLUTION OF MOISTURE CONTENT (A) AND WATER ACTIVITY (B) IN THE MUSCLE *BICEPS FEMORIS* DURING THE PROCESSING OF BAYONNE, PARMA, SERRANO AND IBERIAN HAMS

(Adapted from Monin *et al.* 1997, Vestergaard *et al.* 2000, Toldrá 1991 [unpublished data] and Córdoba *et al.* 1994)

of the product (i.e., composition, direction of the fibers in relation to water movement, etc.) or process conditions (i.e., salting, drying, etc.). In most cases, an apparent or effective diffusivity coefficient, De, has to be estimated. Typical De values at 5C and 10% salt are 3.07×10^{-11} m^2/s in transverse direction to the fibers or 6.11×10^{-11} m^2/s parallel to the fibers (Gou and Comaposada 1997). In fact, there are some factors affecting water diffusion to the outer surface of the ham like pH (low pH favors water loss), amount of intramuscular fat that constitutes a barrier to water diffusion, weight of the ham that extends the time necessary to achieve the desired water loss percentage and water content in the ham. On the other hand, the outer surface is very important because evaporation takes place only on the lean meat area and it is thus necessary to preserve the surface clean and ready for evaporation. The water sorption isotherms are very important to predict the time necessary for drying and estimate the required energy for the drying process.

Salt Diffusion

The movement of salt into the ham is slow. NaCl must diffuse through the liquid phase in muscle, but the heterogeneous structure affects its movement and route. As there is a salt concentration gradient, the diffusion follows the Fick's law. In ideal conditions (in pure water at 25C) the value of D for salt is 1.5×10^{-9} m^2/s (Weast and Astle 1981). The amount of salt in the meat affects its diffusion because of the increase in concentration gradient, but it does not modify the value of D. On the other hand, water moves in the opposite direction (to the outer surface) carrying the Na^+ and Cl^- ions by convection. Thus, an apparent or effective diffusivity coefficient, known as De, has to be estimated. This effective De depends on the pH of meat, type of muscle (especially the presence of intramuscular fat), bones and connective tissue, direction of fibers (parallel or perpendicular), previous freezing/thawing, temperature, etc. Therefore, different values for De such as 0.22×10^{-9} m^2/s for green hams or 0.29×10^{-9} m^2/s for frozen/thawed hams have been reported in the literature (González-Méndez *et al.* 1985). The De value for water is 20 times lower than for salt in meat products. Salt gradient is more important than water gradient so that transportation by convection is very low in comparison to diffusion. The evolution of salt content in the inner and outer muscles of the ham along the processing is shown in Fig. 3.13. After salting, the amount of salt is very high in the outer muscles but low inside the ham. Toward 4–5 months of process, salt tends to equalize, but this profile changes to a reverse profile (higher concentration in the deep muscles) toward the end of the process. This is due to the higher moisture content in those muscles that keeps more dissolved salt (Arnau *et al.* 1995; Monin *et al.* 1997). Salt diffusion through adipose tissue is very low in comparison to the lean meat.

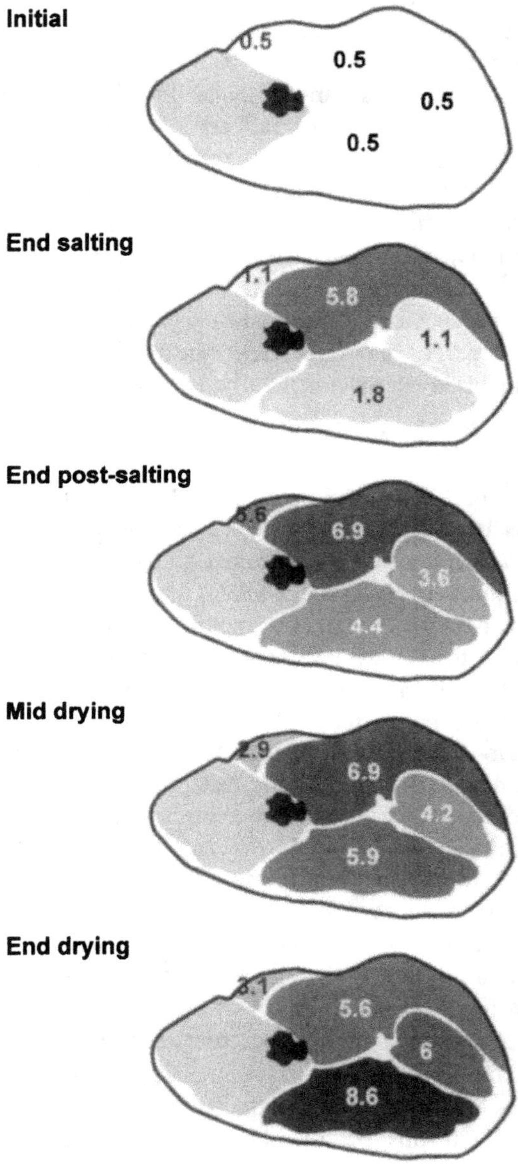

FIG. 3.13. EXAMPLE OF THE EVOLUTION OF SALT IN DIFFERENT INNER AND OUTER MUSCLES ALONG THE PROCESSING OF DRY-CURED HAM
Salt content is expressed as a percentage given in number
(Adapted from Arnau *et al.* 1995 and Toldrá 1991, unpublished data).

Color

The color of meat and meat products is influenced by its moisture and fat content and also by the content of hemoprotein, particularly myoglobin, and its relationship with the environment surrounding it. The color of the dry-cured meat products depends on the concentration of pigment (myoglobin), the degree of conversion to the nitrosyl pigment (usually 10–40%) and the state of the protein (case of denaturation by heat treatment).

Myoglobin is a monomeric, globular heme protein with a molecular mass around 18,000. The myoglobin concentration in pork muscle is between 4 and 9 mg/g. Furthermore, the myoglobin concentration also depends on the type of muscle (Aristoy and Toldrá 1998), being higher in muscles with oxidative patterns than in muscles with glycolytic patterns like M. *Semimembranosus* (about 4.2 mg/g). Most of the muscles in ham exhibit glycolytic or intermediate metabolism (Laborde *et al.* 1985).

The age of the animal also has an important contribution to the increase in the concentration of myoglobin. This is the reason why Iberian hams produced from older Iberian pigs present a darker color than hams produced from standard and younger white pigs. The color may be pinky-red in those products without added nitrate/nitrite or a typical bright-red color from the action of nitrite with myoglobin that will depend on the percentage of myoglobin transformed into nitrosomyoglobin. Some brown or green pigmented areas may appear as a consequence of oxidation by certain peroxide-producing bacteria. The product needs to be hygienically handled to avoid microbial contamination and cut surfaces (especially when sliced) protected from external light. Smoked meat products develop surface colors resulting from the pyrolytic decomposition of wood.

Textural Properties

Texture characteristics of dry-cured ham constitute one of the main sensory attributes perceived by the consumers. Texture mainly depends on the moisture content of the ham, the extent of proteolysis, the content in connective tissue, fat, etc. The enzymatic breakdown of myofibrillar proteins produces an increase in tenderness as discussed in Chap. 6. The proteolysis index gives an indication of the proteases action. Instrumental texture profile analysis has been used to asses textural properties of dry-cured hams, having 12 months and the same moisture content, from exudative, normal and DFD initial quality classes (Tabilo *et al.* 1999). The study of five textural parameters (hardness, springiness, cohesiveness, adhesiveness and chewiness) and resistance to penetration (as an index of hardness during mastication) revealed that all of them were significantly lower in PSE hams than in normal ones. This indicated a higher degradation of the myofibrillar structure and thus a more intense proteolysis in the former group of hams. The electrophoretic analysis of the proteins revealed the full disappearance of titin, nebulin and troponin T and partial of the myosin heavy chain and α-actinin and the appearance of several fragments of 150, 85, 40 and 14.4 KDa. The myosin heavy chain and the fragments of 150 and 85 KDa disappeared in PSE hams (Tabilo *et al.* 1999). The effect of sex revealed a lower moisture content and higher values for the textural parameters in entire males than females.

The content in intramuscular fat has shown a positive influence on some texture and appearance traits of Iberian ham such as oiliness, brightness, juiciness and marbling but is negatively related to dryness, fibrousness and hardness (Ruiz *et al.* 2000).

pH

The pH experiences a slight increase in the different types of hams. It starts at 5.6–5.8, increasing to about 6.1 during the processing, especially during post-salting and initial ripening periods (Fig. 3.14). This increase in pH is related to the generation of free amino acids as a result of proteolysis. The pH remains almost unchanged during the rest of drying/ripening period (Bellatti *et al.* 1985). The evolution of pH in PSE hams has been reported to be similar to the normal hams (Arnau *et al.* 1995). The variations in pH affect positively or negatively the action of the muscle enzymes. For instance, proteolysis has been observed to be more intense in low pH hams, especially toward the end of processing (Buscailhon *et al.* 1994). In addition, pH may affect the substrates of proteolysis since a low pH would alter the myofibrillar structure and make the proteins more accessible to proteases.

Chemical Changes

Several chemical changes are observed along the different stages of the process. They mainly consist of chemical composition as affected by water loss and salt penetration. Other observed changes in the protein and fat fractions are the result of proteases' and lipases' actions and the respective accumulation of protein and fat breakdown products. The oxidation of mono- and polyunsaturated fatty acids and flavor development are described in Chap. 7.

Enzymatic Reactions

There are many biochemical changes along the processing of dry-cured ham as a consequence of enzymatic reactions. Those related to proteolysis/lipolysis and oxidation are described in Chap. 6 and 7, respectively. Proteolytic and lipolytic enzymes are generally stable, being active in most of the cases throughout the whole process. Their action is enhanced when temperature is increased during the drying/ripening stage, and the resulting hydrolytic action is more pronounced (Toldrá 1998). Other muscle enzymes like those involved in postmortem metabolism are mainly active during the first days of process. This is the case for enzymes involved in ATP degradation and the generation of nucleotide breakdown products or the glycolysis-related enzymes and generation of lactic acid as an end product.

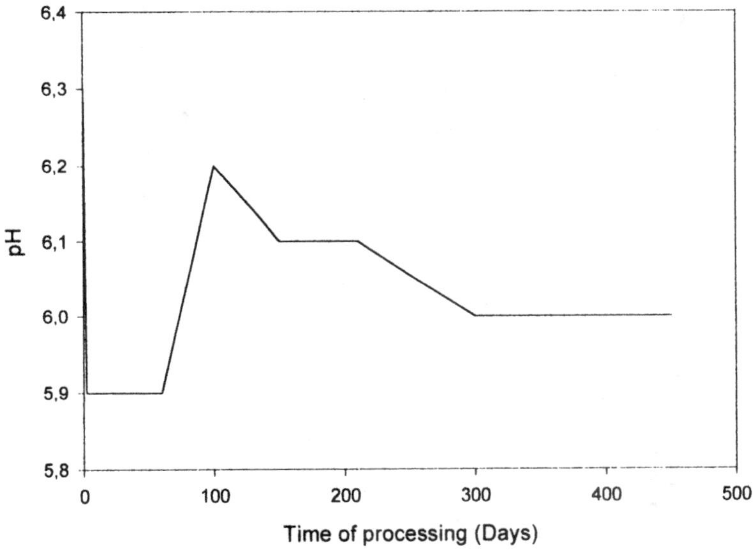

FIG. 3.14. EXAMPLE OF EVOLUTION OF pH DURING THE PROCESSING OF SERRANO DRY-CURED HAM
(Toldrá 1991, unpublished data)

Main Types of Products

The main characteristics of some of the most typical dry-cured hams are compiled in Table 3.3. Most of them depend on the specific hams used as raw material and the conditions that followed for processing as described above. Some further characteristics are briefly described.

The Iberian hams are produced from an autochthonous pig fed and fattened with acorn (mainly from *Quercus ilex L.*, *Q. lusitanica* or *Q. suber*) which is found in the Southwest region of Spain (Fig. 3.15). These pigs are slaughtered at older ages (18–24 months old, 160 kg liveweight) than standard white pigs, favoring the accumulation of fat (better marbling and juiciness) and myoglobin (better color). A cross section of this ham shows a high degree of marbling,

TABLE 3.3.
CHARACTERISTICS OF SERRANO, IBERIAN, PARMA, SAN DANIELE, FRENCH, BAYONNE AND COUNTRY-STYLE DRY-CURED HAMS

Characteristics	**Serrano**[1]	**Iberian**[2]	**Parma**[3]	**San Daniele**[3]	**Bayonne**[4]	**Country-style**[5]
Mean weight (Kg)	10.1	8.0	9.4	9.7	6.9	9.3
Time (Months)	12	24	15	14	9	3
Moisture (%)	48.5	49.0	61.8	60.4	57	64.0
Protein (%)	33.1	24.6	26.9	27.6	30	24.8
Fat (%)	5.9	20.5	3.5	3.6	5	5.3
Salt (%)	8.7	6.5	6.0	6.5	6.2	4.7
Proteolysis index (%)	27.0	-	29.4	28.3	-	-
a_w	0.82	-	0.92	-	0.89	-

(Adapted from [1]Toldrá *et al.* 1997, [2] León-Crespo *et al.* 1986, [3]Baldini *et al.* 1992, [4]Lanore [personal communicaton], [5]Eakes *et al.* 1975)

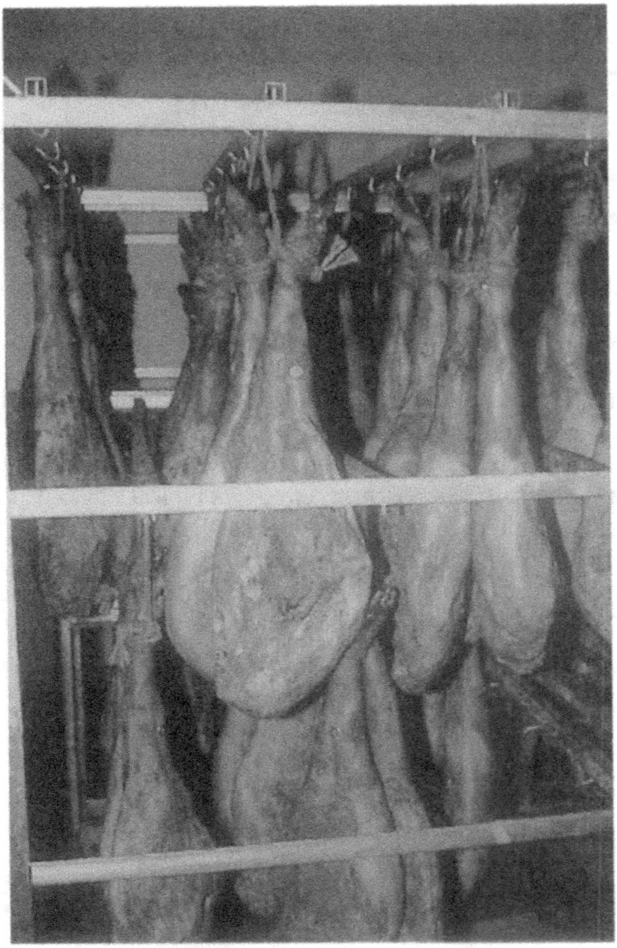

FIG. 3.15. PICTURE OF TYPICAL SPANISH IBERIAN HAMS IN A DRY-CURING CHAMBER

good color, firm texture and exquisite typical flavor (Fig. 3.16). The total processing time may last two or even more years. Corsican hams are produced in Corsica from an autochthonous pig in a sylvo-pastoral extensive system, and they are fattened with chestnuts. These hams are traditionally salted for 3–4 days per kg of ham resulting in hams with a high salt content (up to 11%) and a strong salty taste after 18 months of processing. The final appearance is somewhat similar to the Iberian ham.

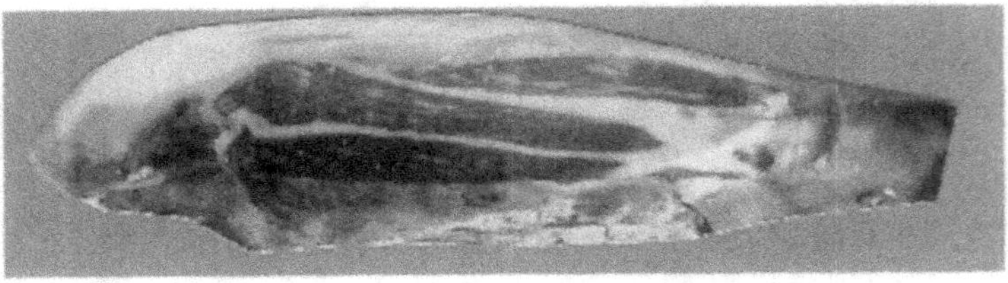

FIG. 3.16. PICTURE OF THE CROSS-SECTION OF A SPANISH IBERIAN DRY-CURED HAM SHOWING AN INTENSE MARBLING

Spanish Serrano and French Bayonne dry-cured hams (Fig. 3.17 and 3.18, respectively) are produced from different crossbreedings of white pigs slaughtered at 6–7 months old (around 110 kg liveweight) and present a cross section with lower marbling, a firm texture and a typical flavor that can be more or less intense depending on the length of the ripening. Heavier pigs (150 kg liveweight) are used in the case of Parma hams (Fig. 3.19). These hams are processed for at least 12 months and no nitrate/nitrite is added since it was banned in 1993 by the specific regulations of the Consortium.

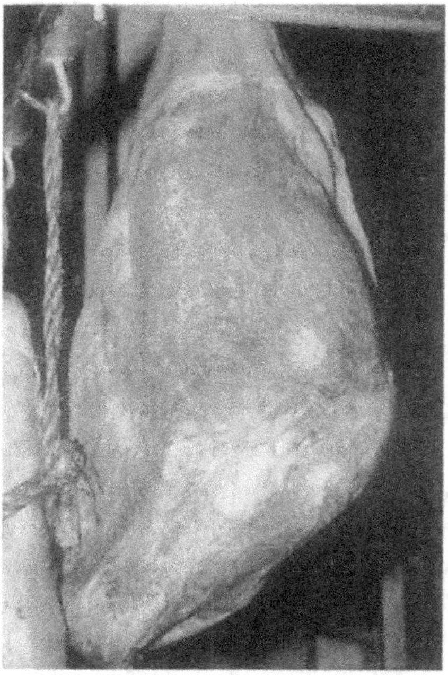

FIG. 3.17. PICTURE OF A TYPICAL SPANISH SERRANO DRY-CURED HAM

FIG. 3.18. PICTURE OF A TYPICAL FRENCH BAYONNE DRY-CURED HAM

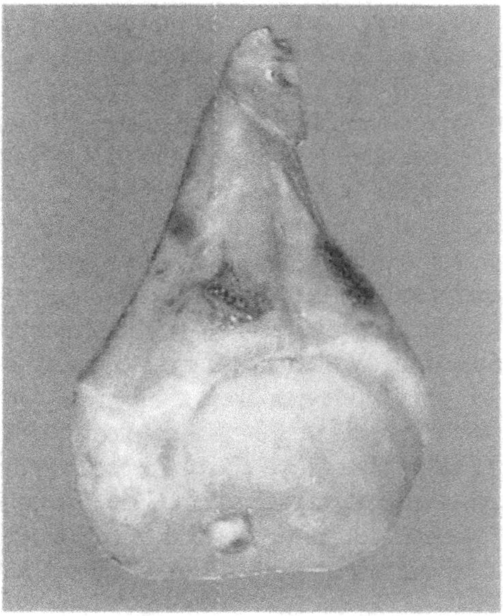

FIG. 3.19. PICTURE OF A TYPICAL ITALIAN PARMA DRY-CURED HAM

Other dry-cured hams are produced in Mediterranean countries from different crossbreedings and with different denominations depending on the local area, specific regulations, etc. Manufacturers of high-quality hams with specific designations (branded hams) must accomplish certain requirements established by respective Consortiums. This is the case of Italian Parma Ham Consortium Regulation (since 1970), Spanish Iberian Guijuelo (since 1986), Spanish Serrano Ham Consortium (since 1999), etc. In all of these cases, there is a specific delimitation of pig production, allowed crossbreeds, feed composition, minimum slaughter age, slaughterhouses and processing technology (i.e., minimum processing time).

Further requirements have been added as scientific knowledge has progressed. For instance, in 1970 the Parma Ham Consortium established criteria for the supply of raw fresh legs and the rejection of defective ones, but, in 1996, new criteria was established for the 12-month Parma ham. These criteria included new requirements for fresh fat quality and pig feeding and new limitations for the percentages of moisture, salt and nonprotein-nitrogen in defatted *Biceps femoris* muscle.

Recently, the European Commission opened a Registry to include determined food products as Protected Designation of Origin (PDO), case of Italian Parma ham, Spanish Guijuelo and Teruel hams, Protected Geographical Indication (PGI), case of French Bayonne hams or Traditional Speciality Guaranteed (TSG), case of Spanish Serrano hams. The number of incorporated products to this Registry is in direct relation to the existing tradition of elaboration or to the trust generated by this protection system.

Country-style ham is produced throughout the United States, but mainly in the southeast (Kentucky and Virginia). It was introduced by settlers who cured them in farms for use in the summer. They used either salt, salt and sugar, with the addition of potassium nitrate, pepper and other spices and smoked the hams over a hickory fire. In recent times, country-style ham is produced at a large scale by private companies. A minimum of 18% of its initial fresh weight and at least 70 days are required for a typical country-cured ham (Fig. 3.20). In some cases, the hams may be aged up to one year to give the country-ham flavor. Most hams are further processed for the retail market.

The traditional German Westphalian ham is smoked on beechwood. The German cold smoked ham (Katenschinken) follows a particular process. A mixture of salt, nitrate, sugar and seasoning is applied by hand rubbing and left to stand in vats for 4 to 9 weeks at 2–4C. Once the excess of dry-curing mixture is removed, the hams are placed on shelves at 8C and low relative humidity for 3 to 4 weeks, then washed with cold water and dried for 12 h. Hams are then rubbed again with seasoning and smoked 3 to 5 hours per day during 4 to 6 weeks (Puolanne 1982).

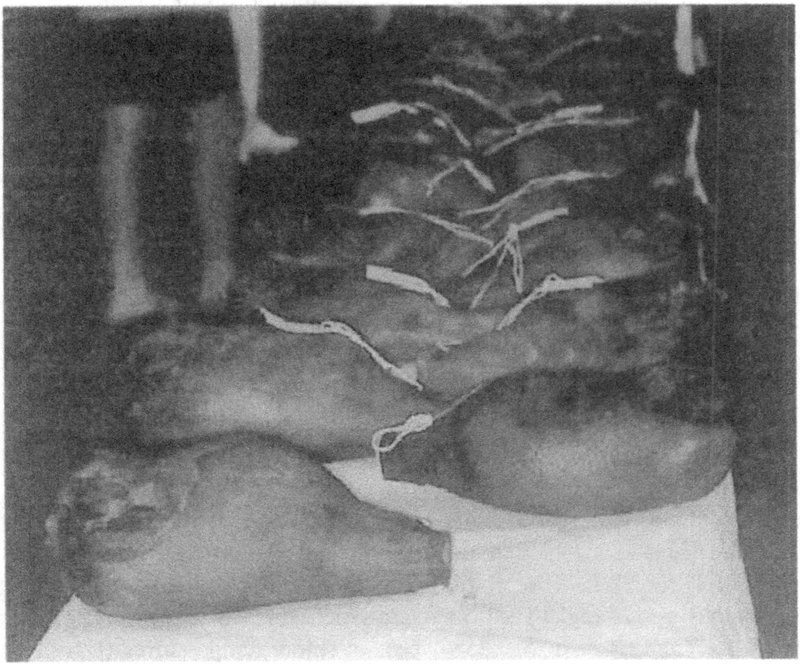

FIG. 3.20. PICTURE OF A TYPICAL AMERICAN COUNTRY-STYLE DRY-CURED HAM
(Courtesy of Dr. W.B. Mikel, University of Kentucky)

The traditional Finnish "sauna" hams are dry-salted and placed in wooden barrels for 1 to 4 weeks. The saunas, typical places in northern Europe for dry-heated baths for the whole family, were originally used as chambers for heating and smoking the hams for several days. Today, the modern industry uses boneless hams that are pickle-injected with multineedle injectors, vacuum tumbled and hot smoked (Puolanne 1982). Some hams like Ching Hua or Yunnan are also produced in China (Campbell-Platt 1995).

Other Dry-cured Meat Pieces: Loins and Shoulders

Dry-cured pork shoulder is processed using the forelegs as raw material. The process is very similar to the dry-curing of ham although it is shorter in all the stages due to the reduced size of the shoulder. Some typical products, like dry-cured lacón, use the forehands. In this case, the process consists of salting for 4–7 days, postsalting for a minimum of 7 days and drying/ripening at 12C and 70% relative humidity for about 1.5 months (Marra *et al.* 1999). The product can be eaten either raw or cooked.

Dry-cured pork loins, like Spanish lomo embuchado or Italian Coppa, are made of the intact/whole loin muscle (*Longissimus dorsi*) piece and require a drying/ripening period although not as long as in the case of dry-cured hams. Typically, an entire piece of loin is rubbed with a wet mixture of curing ingredients, mainly salt, nitrate and nitrite and also sugar and spices such as paprika, pepper and garlic. The loins are left in this mixture for at least one week at refrigeration temperatures in order to allow the seasoning mixture to penetrate. Once the salting stage is finished, the loins are brushed to remove the excess of seasoning mixture and stuffed into casings, natural or regenerated collagen. Then, the loins are hung in a drying chamber for salt equalization for one month at 8C and 72–82% of relative humidity (RH). Finally, the loins are dried 15–20 days at 20–22C for the final reduction in water activity and development of the typical flavor (Hernández *et al.* 1999). The end of the process is achieved when the loins reach a 35–45% weight loss.

Dry-cured beef products are usually made from whole joints of beef that are salted, dried for a few weeks/months and usually smoked at 30–40C. A good example of this kind of manufacture is Spanish Cecina, which is produced from part of the leg and salted with coarse salt and sodium nitrite (at 3–4C and 85–90% RH) for 72 h. The product is washed and postsalted for 30 days and then smoked at 12–15C. The joints are dried for 1.5 months (12–15C and 75–80% RH) and ripened for a few weeks at colder temperatures (García *et al.* 1997). Another example is the Filet d'Anvers of Belgian origin, produced from beef loin and salted (dry or wet) for a few days at 5C. Then, it is heated at 12C for 4–6 days and smoked. Other beef products include the Italian bresaola, the Hungarian pastrami, also produced in the U.S., the Brazilian charqui and the Mexican carne seca.

Trends in Accelerated Processing of Dry-Cured Hams

The dry-curing process is expensive because it needs a long ripening and drying time, which requires buildings for curing chambers under controlled conditions, energy and labor. Thus, any alternative to reduce the time for processing, without affecting quality or safety, is attractive from an economical point of view. Many studies have been carried out to try to accelerate salting penetration into the hams or enhance proteolysis and lipolysis for flavor improvement.

Several techniques have been assayed for accelerating the dry-curing process, most of them in country-style hams (Marriott *et al.* 1992). Some of the first assayed techniques consisted of the application of enzymes such as papain to fresh hams by injection into the femoral artery (Smallings *et al.* 1972). Even though the treated hams were more tender, the quality was considered objectionable. Boning and skinning of hams had the advantage of a rapid weight loss, which is considered desirable for dry-cured hams, and a higher tenderness and concentration of salt, but with a slightly lower overall satisfaction as compared to the unskinned hams (Montgomery *et al.* 1976; Marriott *et al.* 1983). The mechanical tenderization through blade penetration prior to dry-cure application produced a minimal effect on the process, achieving only a higher percentage of salt (Marriott *et al.* 1985). On the other hand, tumbling (which consists of a revolving drum with baffles) increases cure penetration and migration of NaCl and other cure adjuncts through the ham with good distribution and uniformity. It also increases water loss but exerts physical damages which lead to increased saltiness, moisture loss and physical distortion without improving any quality characteristic in the final product (Leak *et al.* 1984).

The direct use of nitric oxide instead of nitrate or nitrite is another approach to accelerate curing time, especially color development, although further studies are necessary (Marriott *et al.* 1992). The microbial flora of the hams has been studied because microorganisms could be used to accelerate flavor development. However, use of microbial inoculation has not resulted in sensible improvements in the sensory characteristics (Marriott *et al.* 1987) and no further real applications have been developed in view of this minimal effect.

Pre-freezing of hams produces a slight acceleration of proteolysis and lipolysis, especially around 3–6 months for Spanish dry-cured ham reaching the end of the process. These hams have similar sensory scores, except for a higher salty taste, when compared to standard hams (Motilva *et al.* 1994). However, it should be taken into account that the uptake of salt is higher in frozen/thawed hams so that it is necessary to reduce the time for salting by a couple of days.

Finally, the use of specific enzymes constitutes an alternative and attractive way to accelerate the process that is currently under study since much more information is available on the role and characteristics of the endogenous muscle enzymes (Toldrá and Flores 1998). So, there are two main ways: steering the adequate conditions into the ham for an optimal and selective *in situ* action of muscle proteases and lipases or adding the enzymes (i.e., microencapsulated in liposomes) by rubbing them onto the outer surface of the ham with the cure for their penetration into the ham, then diffusion during the salting/post-salting stages and action on the ripening stage when the temperature increases. Care should be taken with the temperature since an excessive increase in temperature, even though it may apparently accelerate the process because of the activation of the enzymes, can result in an excess of dryness on the ham surface. This can negatively impact water diffusion to the surface and evaporation.

REFERENCES

ARISTOY, M-C. and TOLDRÁ, F. 1998. Concentration of free amino acids and dipeptides in porcine skeletal muscles with different oxidative patterns. Meat Sci. *50*, 327–332.

ARNAU, J., GUERRERO, L., CASADEMONT, G. and GOU, P. 1995. Physical and chemical changes in different zones of normal and PSE dry cured ham during processing. Food Chem. *52*, 63–69.

BALDINI, P., BELLATTI, M., CAMORALI, G., PALMIA, F., PAROLARI, G., REVERBERI, M., PEZZANI, G., GUERRIERI, C., RACZYNSKI, R. and RIVALDI, P. 1992. Characterization of Italian raw ham by chemical, physical, microbiological and organoleptical parameters. Ind. Conserve *67*, 149–159.

BELLATTI, M., DAZZI, G., CHIZZOLINI, R., PALMIA, F. and PAROLARI, G. 1985. Modifications chimiques et physiques des proteins au cours de la maturation du Jambon de Parme. Viandes Produits Carneés *6*, 142–145.

BEM, Z., ZLENDER, B. and HECHELMAN, H. 1995. Desirable and undesirable effects of smoking meat products. Die Fleischerei *3*, 3–8.

BUSCAILHON, S., BERDAGUE, J.L., GANDEMER, G., TOURAILLE, C. and MONIN, G. 1994. Effects of initial pH on compositional changes and sensory traits of French dry-cured hams. J. Muscle Foods *5*, 257–270.

CAMPBELL-PLATT, G. 1995. Fermented meats-A world perspective. *In* Fermented Meats, G. Campbell-Platt and P.E. Cook (eds.), pp. 39–52. Blackie Academic & Professional, London.

CASSENS, R.G. 1995. Use of sodium nitrite in cured meats today. Food Technol. *49*, 72–81.

COMAPOSADA, J., GOU, P. and ARNAU, J. 2000. The effect of sodium chloride content and temperature on pork meat isotherms. Meat Sci. *55*, 291–295.

CÓRDOBA, J.J., ANTEQUERA, T., GARCÍA, C., VENTANAS, J. LÓPEZ-BOTE, C. and ASENSIO, M.A. 1994. Evolution of free amino acids and amines during ripening of Iberian cured ham. J. Agric. Food Chem. *42*, 2296–2301.

EAKES, B.D., BLUMER, T.N. and MONROE, R.J. 1975. Effect of nitrate and nitrite on color and flavor of country-style hams. J. Food Sci. *40*, 973–976.

EUROPEAN DIRECTIVE (1995) Additives different from colorants and edulcorants used in foods. Ref. 95/2/CE of 20 February. European Commission. Brussels, Belgium.

FRENTZ, J-C. and ZERT, P. 1990. L´encyclopédie de la charcuterie. Dictionnaire encyclopédique de la charcuterie. 3rd Ed., pp. 435. Soussana, Orly, Paris.

GARCIA, I., DIEZ, V. and ZUMALACARREGUI, J.M. 1997. Changes in proteins during the ripening of Spanish dried beef Cecina. Meat Sci. *46*, 379–385.

GONZALEZ-MENDEZ, N.F., GROS, J.B., POMA, J.P. and RAMOS, E. 1985. Influencia de la congelación sobre la difusión del cloruro sódico en el músculo *Longissimus dorsi* del puerco. Rev. Agroquim. Tecnol. Aliment. *25*, 279–286.

GOU, P. and COMAPOSADA, J. 1997. Water transfer inside the dry-cured ham during the drying process. Eurocarne *58*, 33–39.

HERNANDEZ, P., NAVARRO, J.L. and TOLDRÁ, F. 1999. Lipolytic and oxidative changes in two Spanish pork loin products: dry-cured loin and pickled-cured loin. Meat Sci. *51*, 123–128.

KEMP, J.D., ABIDOYE, D.F.O., LANGLOIS, B.E., FRANKLIN, J.B. and FOX, J.D. 1980. Effect of curing ingredients, skinning, and boning on yield, quality, and microflora of country hams. J. Food Sci. *45*, 174–177.

KEMP, J.D., LANGLOIS, B.E. and JOHNSON, A.E. 1982. Effect of pre-cure freezing and thawing on the microflora, fat characteristics and palatability of dry-cured ham. J. Food Prot. *45*, 244.

LABORDE, D., TALMANT, A. and MONIN, G. 1985. Activités enzymatiques métaboliques et contractiles de 30 muscles du porc. Relations avec le pH ultime atteint après la mort. Reprod. Nutr. Develop. *25*, 619–628.

LEAK, F.W., KEMP, J.D., LANGLOIS, B.E. and FOX, J.D. 1984. Effect of tumbling and tumbling time on quality and microflora of dry-cured hams. J. Food Sci. *49*, 695–698.

LEÓN-CRESPO, F., MARTINS, C., PENEDO, J.C., BARRANCO, A., MATA, C., and BELTRÁN, F. 1986. Diferencias en la composición química de ocho regiones anatómicas del jamón serrano Ibérico. Alimentaria *23*, 23–27.

MARRA, A.I., SALGADO, A., PRIETO, B. and CARBALLO, J. 1999. Biochemical characteristics of dry-cured lacón. Food Chem. 67, 33–37.

MARRIOTT, N.G., GRAHAM, P.P. and CLAUS, J.R. 1992. Accelerated dry curing of pork legs (hams): a review. J. Muscle Foods *3*, 159–168.

MARRIOTT, N.G., GRAHAM, P.P., SHAFFER, C.K. and PHELPS, S.K. 1987. Accelerated production of dry cured hams. Meat Sci. *19*, 53–64

MARRIOTT, N.G., KELLY, R.F., SHAFFER, C.K., GRAHAM, P.P. and BOLING, J.W. 1985. Accelerated dry curing of hams. Meat Sci. *15*, 51–62

MARRIOTT, N.G., TRACY, J.B., KELLY, R.F. and GRAHAM, P.P. 1983. Accelerated processing of boneless hams to dry-cured state. J. Food Prot. *46*, 717–721.

MONIN, G., MARINOVA, P., TALMANT, A., MARTIN, J.F., CORNET, M., LANORE, D. and GRASSO, F. 1997. Chemical and structural changes in dry-cured hams (Bayonne hams) during processing and effects of the dehairing technique. Meat Sci. *47*, 29–46.

MONTGOMERY, R.E., KEMP, J.D. and FOX, J.D. 1976. Shrinkage, palatability, and chemical characteristics of dry-cured country ham as affected by skinning procedure. J. Food Sci. *41*, 1110–1115.

MOTILVA, M-J., TOLDRÁ, F., NADAL, M-I. and FLORES, J. 1994. Prefreezing hams affects hydrolysis during dry-curing. J. Food Sci. *59*, 303–305.

PAROLARI, G. 1996. Review: Achievements, needs and perspectives in dry-cured ham technology: the example of Parma ham. Food Sci. Tech. Int. *2*, 69–78.

PUOLANNE, E. 1982. Dry cured hams — European style. Proc. Reciprocal Meat Conf. *35*, 49–52. Blacksburg, Virginia.

RUIZ, J., VENTANAS, J., CAVA, R. ANDRÉS, A.I. and GARCÍA, C. 2000. Texture and appearance of dry cured hams as affected by fat content and fatty acid composition. Food Res. Int. *33*, 91–95.

SMALLINGS, J.B., KEMP, J.D., FOX, J.D. and MOODY, W.G. 1992. Effect of antemortem injection of papain on the tenderness and quality of dry-cured hams. J. Anim. Sci. *32*, 1107–1112.

SPANIER, A.M., FLORES, M. and TOLDRÁ, F. 1999. Flavor differences due to processing in dry-cured and other ham products using conducting polymers (electronic nose). *In* Flavor Chemistry of Ethnic Foods, F. Shahidi and C-T. Ho (eds.), pp. 169–183. Kluwer Academic/Plenum Publ., New York.

TABILO, G., FLORES, M., FISZMAN, S.M. and TOLDRÁ, F. 1999. Postmortem meat quality and sex affect textural properties and protein breakdown of dry-cured ham. Meat Sci. *51*, 255–260.

TOLDRÁ, F. 1992. The enzymology of dry-curing of meat products. *In* New Technologies for Meat and Meat Products. J.M. Smulders, F. Toldrá, J. Flores and M. Prieto (eds.), pp. 209–231. Audet, Nijmegen, The Netherlands.

TOLDRÁ, F. 1998. Proteolysis and lipolysis in flavour development of dry-cured meat products. Meat Sci. *49*, S101–110.

TOLDRÁ, F. and FLORES, M. 1998. The role of muscle proteases and lipases in flavor development during the processing of dry-cured ham. CRC Crit. Rev. Food Sci. Nutr. *38*, 331–352.

TOLDRÁ, F., FLORES, M., NAVARRO, J-L., ARISTOY, M-C. and FLORES, J. 1997. New developments in dry-cured ham. *In* Chemistry of Novel Foods, H. Okai, O. Mills, A.M. Spanier, and M. Tamura (eds.), pp. 259–272. Allured Pub. Co., Carol Stream, IL.

CHAPTER 4

PRINCIPLES OF DRY-FERMENTED SAUSAGE-MAKING

There are many types of dry-fermented sausages with specific characteristics that are typical of particular geographical locations with different eating habits, history, tradition, religion and climate. These products, having a great variety of flavors, colors and texture, constitute an important part of the local economy, particularly in its culture and gastronomic heritage. It is very difficult to classify dry-fermented sausages because of the high number of different products. The main classification is usually based on the extent of drying (weight loss) and/or time for ripening (Table 4.1).

TABLE 4.1.
EXAMPLE OF CLASSIFICATION OF SOME FERMENTED SAUSAGES

Product type	Weight loss (%)	Ripening	Denomination
Semi dry-fermented	< 20	Rapid	Summer sausage
" "	"	"	Saucisson d´Alsace
Dry-fermented	> 30	Slow	Italian salami
"	"	"	Spanish salchichón
"	"	"	French saucisson
"	"	"	Pepperoni
"	"	"	Spanish chorizo
"	"	Regular	Turkish-style soudjouk

Taking into account the moisture content, the products may be classified as semi-dry (weight loss lower than 20%) or dry (weight loss higher than 30%) fermented sausages. The length of the ripening period will vary depending on the kind of product, the desired final quality and its diameter. The total processing time may take less than 7 days (rapid process), around 3 weeks (regular process), or 90–120 days or even up to 150–180 days (slow process). The sensory properties will be strongly influenced by the length and conditions of the process as well as the optional smoking (Flores 1996). A diagram containing all the stages of the process, from the reception of ingredients and additives to the final product, is shown in Fig. 4.1. The raw materials, as described below, must strictly meet fixed standards in order to guarantee the final hygiene and sensory quality of the product.

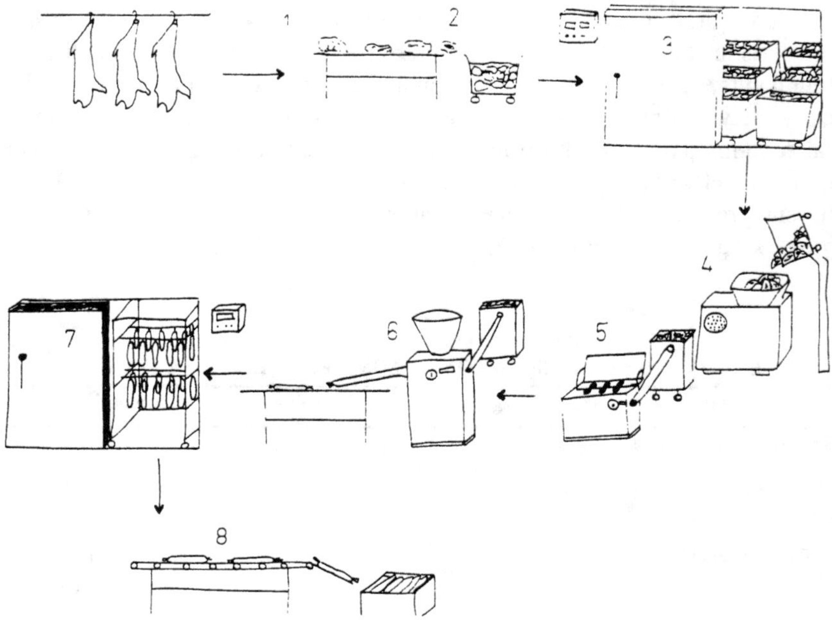

FIG. 4.1. DIAGRAM SHOWING ALL THE STAGES IN THE PROCESSING OF DRY-FERMENTED SAUSAGES. (1) RECEPTION, (2) DEBONING AND CUTTING, (3) COOLING, (4) GRINDING/CHOPPING, (5) VACUUM TUMBLING, (6) STUFFING, (7) FERMENTATION AND RIPENING, (8) PACKAGING

INGREDIENTS AND ADDITIVES

Meat

Sausages are usually made from lean pork, mixtures of pork and beef, or solely beef. Typical proportions are 50–60% lean pork and 10–20% beef. Some other animal species may be used depending on the type of product and geographic location. The selection of adequate meat cuts and trimmings, which are kept under strict preservation conditions and have especially good microbiological and chemical characteristics, is extremely important. It is necessary to perform bacteriological controls and check the temperature and pH of meats at reception. Pork meat with pH values between 5.6 and 6.0 should be chosen (Roca and Incze 1990). DFD meat should be avoided because of its high water binding and neutral pH that favors spoilage. On the other hand, PSE meats may cause serious color problems (paleness) and protein denaturation, although some PSE meats may be used under controlled conditions without affecting the final quality. The meat from older animals, which contains a higher amount of myoglobin, is better because of the intensity of the cured color. The use of beef combined with pork meat has some advantages, such as color improvement (higher content of myoglobin), a better protein solubilization and an improved drying rate. Beef meat with a high pH should also be rejected for the same safety reasons as pork.

Fat

It is recommended to use fresh fats or at least fats that have not been stored for a long time. Endogenous lipases are active at low temperatures and some lipolysis (generation of free fatty acids) is observed during cold storage and even after several months of frozen storage. The amount of unsaturated fatty acids is also important because they are prone to oxidation and therefore can develop off-flavors and rancidity. Firm pork back fat, which has a high melting point (low unsaturation level), is the most adequate (Lücke 1985) in order to avoid adverse effects on flavor (rancidity) during processing. In some cases, such as in Muslim countries where pork is not consumed for religious reasons, other fats may be used. For instance, fat-tailed sheep are used in Turkish-style sausages (soudjouk). Periodic controls of quality like iodine index (as an indicator of unsaturation) and acid index (as an indicator of freshness) are recommended. Soft fats, which usually contain a high percentage of polyunsaturated fatty acids, must be avoided because they can create problems during chopping and mixing and are prone to develop off-flavors and rancidity.

Curing Agents

Salt, nitrate and nitrite are the main curing agents. Salt is always present in cured meat products and has several roles: a bacteriostatic effect, inhibiting the growth of undesirable microorganisms and allowing lactic acid bacteria to become the predominant group; an influence on flavor since it gives a characteristic salty taste; and an increase of myofibrillar protein (actin and myosin) solubility. The usual amounts of salt in the mix are around 2–3%.

The addition of nitrite, previously formulated in a curing salt or combined with nitrate, is a common practice. Nitrate can be added as sodium or potassium salt and is generally used for long-term ripened sausages. Nitrate is reduced by nitrate reductase activity from bacteria that are either naturally present in the mix or added as starter cultures (such as *Micrococcaceae*). Therefore, nitrite is produced from nitrate or added as sodium or potassium salt as part of the curing salt. The maximum amount allowed for sodium nitrite is 156 ppm (1/4 ounce per 100 pounds of meat) in the U.S. In the case of the European Union, a European Directive (1995) allows maximum and residual amounts of nitrate and nitrite in meat products.

Nitrite is very reactive and interacts with protein and other meat components. It reacts with myoglobin as an essential reaction for the development of the characteristic red cured color, acts as a preservative inhibiting the growth of undesirable microorganisms (with specific protection against *Clostridium botulinum*) and contributes to the typical cured flavor (Cassens 1997). The nitrate and nitrite residues are generally low in the finished product. Due to the potential toxic effects of N-nitroso compounds, the amounts of nitrate and/or nitrite added to the initial mixture are being reduced to those strictly necessary for protection against botulism (Cassens 1995).

Ascorbic and erythorbic acids or their sodium salts are used as reducing agents to speed up the curing reactions. They facilitate the reduction of nitrite to nitric oxide, exert antioxidant activity stabilizing color and flavor and, most important, inhibit the formation of nitrosamines.

Carbohydrates

The glucose content of postrigor meat is very low and does not allow for significant pH reduction (Lücke 1985). The major function of added carbohydrates is to provide a substrate for the biological acidulation by lactic acid bacteria. The rate and extent of lactic acid formation, pH drop and evolution of the microflora will depend on the amount and type of carbohydrates added to the mix. So, if large amounts (e.g., 2%) of a highly metabolizable sugar like glucose are added, the pH will drop very fast and reach values as low as 4.5, inhibiting nitrate reductase and some of the most important enzymes responsible for the generation of flavor compounds. On the other hand, the addition of low amounts or long chain carbohydrates (low metabolizable sugars) may result in a deficient pH reduction that will allow the growth of undesirable microorganisms. In general, the amount of sugar may vary between 0.5 and 1% but may reach 2% in some semi-dry fermented sausages. Glucose and saccharose are metabolized quickly and ensure a rapid acidification. Lactose follows at a slower rate than glucose. Dextrines or starch are metabolized slowly and their use is recommended for long ripening sausages. Lactic acid is the main compound resulting from the stoichiometry of homofermentation. Other compounds like ethanol, acetic acid, CO_2, etc., may appear in minor concentrations through the heterofermentative pathway. Higher amounts may cause flavor or textural problems. For instance, an excess of carbon dioxide gas production can produce pinholes within the sausage and even break the casing by product expansion (Bacus 1984).

In high-quality dry-fermented sausages ripened at mild temperatures for long periods of time, the amount of saccharose or glucose may be as low as 0.1–0.3%. In some artisanal sausages no carbohydrates are added.

Spices and Flavorings

Spices are natural products of plant origin subjected to climatic conditions that may slightly affect their quality. They can be seeds (e.g., mustard), leaves (e.g., oregano, rosemary), bulbs (e.g., garlic, onion) or fruits (e.g., pepper, paprika) and used either in their natural form (whole or ground) or as flavoring extracts (essential oils and oleoresins). Spices are commonly used in dry-fermented sausages for imparting a characteristic and typical flavor, sometimes color too (with paprika), to the product. Some spices have antioxidant properties. There is usually no more than 1% spice added, although ground pepper, paprika and mace are some of the spices with more extended use and application. Care must be taken with microbial contamination, since spices may contain high numbers of bacteria, which would make it necessary to correctly control and disinfect.

Flavoring agents and flavor enhancers may be used to accentuate a specific flavor. Smoke flavoring may be applied on the surface as an oil or water solution to give a smoke flavor. Monosodium glutamate is used to enhance the flavor of the final product. Its enhancing action may be potentiated in the presence of small amounts of inosine monophosphate (5´-IMP) or guanosine monophosphate (5´-GMP), naturally present in the meat or added as an additive.

STARTER CULTURES

The use of starter cultures has expanded in recent years because it allows for safe production and more desirable and uniform product characteristics through the different batches. The use and properties of starter cultures are described in Chap. 5.

CASINGS

Casings may be of natural or synthetic origin. Animal cases are typical in traditional sausages. They are irregular, lacking in uniformity, but give a handmade appearance. These cases are usually stored in dry salt and need extensive washing and disinfecting, since they can be contaminated, prior to their use. Synthetic cases are made of materials permeable to evaporation gas and smoking (e.g., restructured collagen). They have important advantages such as a high degree of uniformity, a regular pore size that allows good control drying and they are available in any diameter required. It is important for casing to follow the shrinkage of the product during drying to avoid loosening from the meat.

PROCESSING TECHNOLOGY

Once the ingredients, additives and starter cultures are received and their quality verified, the next stage involves their combination initiating the processing. A flow diagram with the simulation of the processing of dry-fermented sausages is shown in Fig. 4.2. Different examples of manufacturing processes for several dry and semi-dry fermented sausages are shown in Table 4.2.

Comminution

The chilled meat pieces (usually below -4C although sometimes they may be at -2C to +1C) and frozen fat tissues (below -8C) are chopped and ground in a grinder. It may be performed simultaneously or by separate plates with a latter mixture. It is important to use cold temperatures in order to avoid smearing the fat and the formation of a fine film over the lean meat, which will reduce its ability to lose moisture (Roca and Incze 1990). The size of lean and fat particles will depend on the plate size during grinding. An example of grinder plates and knives is shown in Fig. 4.3. Then, the curing additives (salt, nitrate and/or nitrite), adjuncts (carbohydrates, flavoring agents, spices and optionally sodium ascorbate) and microbial starters are added. The ground mass is homogenized by mixing under a vacuum for removing as much oxygen as possible (Fig. 4.4). The oxygen interferes with the formation of the desired sensory properties such as color and flavor. It is very important to have the mix as homogeneous as possible. The batter is stuffed at 0–1C into casings either natural (mostly for traditional products) or manmade (collagen or synthetic) by using vacuum filling machines (Fig. 4.5) with devices of different diameters (Fig. 4.6). The use of a vacuum avoids the presence of bubbles, by allowing trapped air to escape during stuffing, and disruptions in the casing. The pore size of the casings must allow evaporation of water during fermentation/ripening/drying and penetration of smoke. The rate of drying is affected by casing diameter or sausage size, but not by the length of the sausage or its weight (Acton 1977). Mold growth may be avoided by dipping the sausage into a 2.5% solution of potassium sorbate in water. In some cases, like some Mediterranean dry-fermented sausages, the growth of a mold layer on the outer

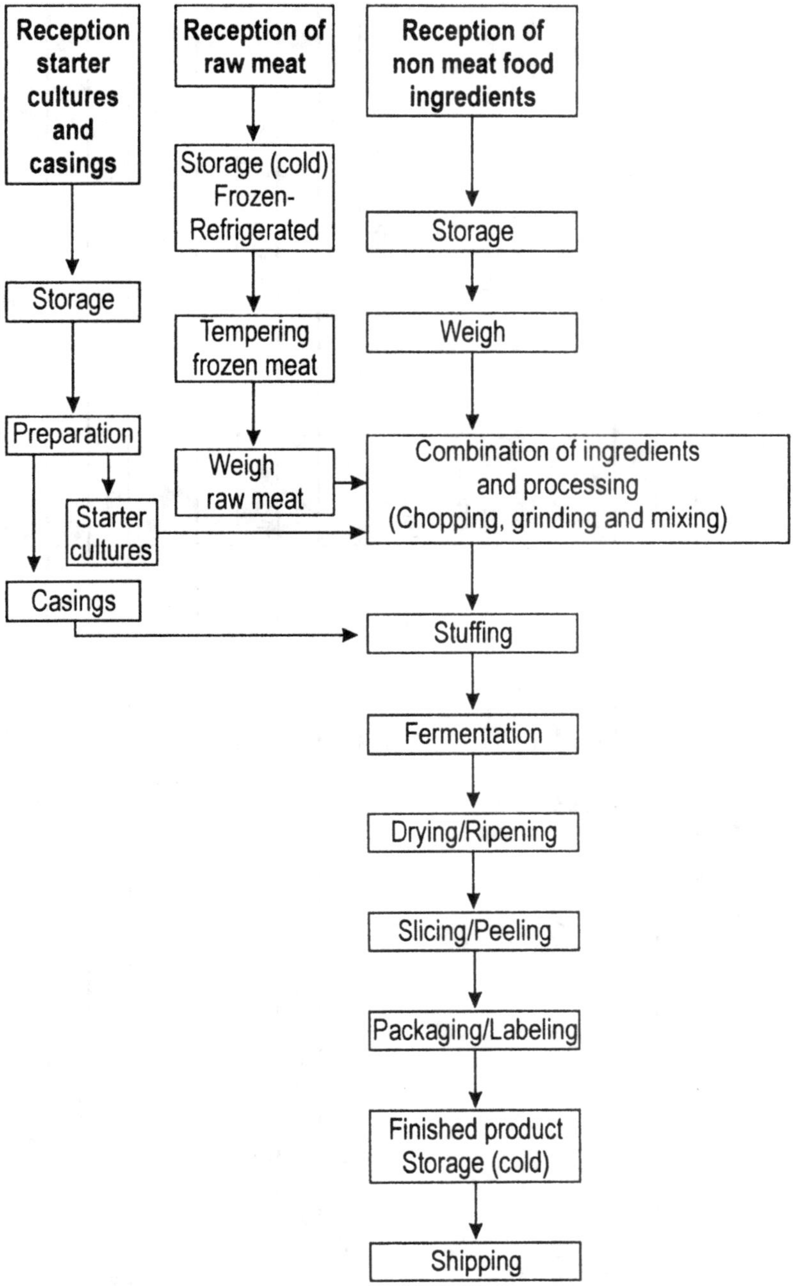

FIG. 4.2. FLOW DIAGRAM FOR THE PROCESSING OF DRY-FERMENTED SAUSAGES

TABLE 4.2.
EXAMPLES OF MANUFACTURING PROCESSES FOR SEVERAL DRY AND SEMI-DRY SAUSAGES

	Summer sausage[a]	**Pepperoni**[b]	**Spanish chorizo**[c]	**Turkish-style soudjouk**[d]
Fat (%)	20-30	17	20	10
Plate hole size (mm)	3	4.6	7	3
Salt (%)	2.5-3.5	3	2.5	2
Sugar (%)	Glucose 2	Sucrose 1, Glucose 1	Glucose 1.8	Sucrose 0.6
Nitrate/nitrite (ppm)	$NaNO_2$ 150/ $NaNO_3$ 150	$NaNO_3$ 400	$NaNO_3$ 300	$NaNO_3$ 330 /$NaNO_2$ 50
Other additives (%)		Crushed red pepper 0.2 Ground cayenne pepper 0.2, Pimento 0.5 Whole anise seed 0.2 Garlic powder 0.01	Sodium ascorbate 500 ppm, Garlic 0.5 Pepper 3 Oregano 0.12	Garlic 1 Red pepper 0.7 Black pepper 0.5 Cumin 0.9 Pimento 0.25 Refined olive oil 0.25
Case diameter (mm)	50-100	55	40	
Resting	3-4°C for 2-4 d	-	24 h at 3-4°C	-
Fermentation: T (°C)	24-32	35	20-24	12-14
" RH (%)	85	85	90	95-85
" t (days)	0.5-2	1-3	1-2	18
Ripening : T (°C)	38-43	12	14-16	
" RH (%)	65-75	65	80	
" t (days)	2	40-42	15-90	

TABLE 4.2. (Continued)
EXAMPLES OF MANUFACTURING PROCESSES FOR SEVERAL DRY AND SEMI-DRY SAUSAGES

	Salami Milano[e]	**Hungarian salami**[f]	**Spanish salchichón**[c]	**Saucisson d´Alsace**[f]
Fat (%)	25	25-30	15-30	35
Plate hole size (mm)	3-4	10	8	6
Salt (%)	3.5	2.6	2.5-3	3
Sugar (%)	6	2	Glucose 0.8, Lactose 1	Glucose 0.3, Lactose 2
Nitrate/nitrite (ppm)	$NaNO_3$ 250	$NaNO_3$ 300	$NaNO_3$ 300	$NaNO_3$ 350
Other additives (%)	White pepper 0.1-0.2 Garlic 0.01	White pepper 0.25 Garlic 0.02 Red wine 0.4 Paprika 0.1	Sodium ascorbate 500 ppm Ground black pepper 0.3	Garlic 0.1 Gray pepper 0.3 Rhum 0.5 Spices 0.05
Case diameter (mm)	60-110	60	40-60	60
Resting	24h at 0°C	5d at 3°C	-	-
Fermentation: T (°C)	20-24	20	20-24	20
" **RH (%)**	80-90	smoking	95-90	smoking
" **t (days)**	0.3-2	4	2	2
Drying: T (°C)	16-20			
" **RH (%)**	65-85			
" **t (days)**	5-7			
Ripening : T (°C)	10-13	10	12-15	13-15
" **RH (%)**	80-85	80-85	80	70-75
" **t (days)**	60-180	120-180	30-200	10-40

([a]Buege and Cassens 1980, [b]Palumbo *et al.* 1976, [c]Flores 1989, [d]Gokalp 1985, [e]DelMonte 1990 and Novelli *et al.* 1998, [f]Frentz and Zert 1990)

FIG. 4.3. GRINDER PLATES AND KNIVES ARE COMMERCIALLY AVAILABLE IN MANY SIZES AND CONFIGURATIONS

FIG. 4.4. VACUUM MIXER MASSAGER. THE DESIGN AND CONFIGURATION OF PADDLES MAY VARY DEPENDING ON COMMERCIAL SUPPLIERS

FIG. 4.5. EQUIPMENT FOR STUFFING UNDER VACUUM

surface of the sausage is typical and can contribute to flavor and appearance by reducing the oxygen levels and subsequent oxidations and increasing the pH by generating ammonia.

Fermentation

The main objectives are to activate microbial development and achieve a correct pH drop. The sausages are hung in maturing rooms or smokehouses (as smoke may be applied during this stage if not inhibiting the fermentation process). These rooms are usually air-conditioned ripening chambers with computerized temperature (T) control, relative humidity (RH) and air velocity.

Natural ripening chambers are traditionally used when producing sausages in an artisanal way. The nature of the fermentation will depend on which microorganisms are present. The inoculation ratio and the processing schedule are selectively altered to optimize the quality of the final product. Mediterranean sausages, which usually undergo a slow process with nitrate addition, require a few hours for nitrate reduction to nitrite by nitrate-reducing bacteria. In other cases, the intense acidification produced at high fermentation temperatures inhibits nitrate-reductase and catalase promoting bacterial growth and may contribute to color defects or changes in the flavor (Flores and Bermell 1996). When nitrate is used, the sausage mixture is usually left to stand around 24 h under refrigeration before the fermentation to allow development and action of nitrate reductase. On the other hand, when only nitrite is used in faster processes (e.g., North-European sausages), they are smoked in most cases (Leistner 1995). In Italy and France it is common to keep the salami under refrigeration (2–5C) for 2–4 days in order to reduce the acidification rate and obtain sausages with a milder sour taste.

FIG. 4.6. EXAMPLES OF STUFFING DEVICES WITH DIFFERENT DIAMETERS

Sausages are equilibrated for a short time (around 6 h) at the fermentation temperature but with a low relative humidity (RH about 60%) to prevent water condensation on the outer surface of the sausage. The conditions can vary widely during fermentation depending on the type of sausage being made. The fermentation may take 12–48 h depending on the temperature, being longer at lower temperatures. There is a good correlation of the temperature and A_w of the meat mixture and bacterial growth. Low fermentation temperatures, around 22–26C, are used in European-style dry-fermented sausages although there are some variations. For instance, the fermentation time usually applied in Italy is very short to increase the product temperature, 2–3 h for salamis with 60 mm diameter and 4–6 h for salamis with 80–90 mm diameter. The relative humidity is high (>80%) during this period in order to increase the heat transfer. Afterward, the drying stage is carried out for a few days at 16–22C and variable relative humidities (successive reductions between 90% and 65%) to allow water diffusion inside the salami. Ripening is carried out at low temperatures (9–13C) and high relative humidities (80–90%) (Baldini 2001, personal communication).

In the U.S., the fermentation temperature is high, around 30–37C, to allow the added starters such as *L. plantarum* or *P. acidilactici* to produce high amounts of lactic acid and achieve pH values below 5.0 to 4.6. This rapid acid production through a fast fermentation ensures the inhibition of spoilage microorganisms. The air velocity is maintained around 0.1–0.5 m/s and the relative humidity around 90%, which keeps a progressive and slow drying but avoids an excessive and undesirable dryness on the outer layer of the sausage. During fermentation the sausages start to develop a pH drop from lactic acid production and accumulation, a characteristic color resulting from nitric oxide reaction with myoglobin, initial drying as a result of the combined action of temperature and relative humidity and consistency because the decline in pH toward the isoelectric point of muscle proteins produces water loss and a sensible decrease in the solubility of sarcoplasmic and myofibrillar proteins.

Chemical Acidulation

An alternative to bacterial fermentation consists of the addition of chemical compounds to produce a pH drop as a simulation of that obtained by fermentation. Chemical acidulation is sometimes used in order to reduce the long time required for bacterial fermentation and thus reduce production costs and obtain a cheaper product. Glucono-delta-lactone is the most commonly used. It is hydrolyzed to gluconic acid and produces a rapid decrease in pH. However, although gluconic acid produces a very fast pH drop, its values are a little bit higher than those obtained with lactic acid fermentation. Of course, the final quality of the product is rather poor because the rapid pH drop implies a drastic reduction in endogenous and bacterial enzyme activity, especially in enzymes like exo-peptidases and lipases closely related with flavor development. The result is thus a lack of flavor in the final product.

Smoking

Smoking is also one of the oldest processes used in the preservation of food stuffs. It may be applied during fermentation or during the ripening at regular intervals. A smoke solution is formed when the vapor phase of smoking dissolves in the surface and interstitial water of the meat mixture. It can also be applied, sprayed or atomized on the surface of the sausage, as a liquid produced by distillation and subsequent condensation of volatile compounds from smoke. The application of smoking is optional and depends on the type of product and the country. It is more common in northern countries where the application of ripening/drying is more difficult because of the climate. It can be accompanied by heating up to 60C. Smoking has several advantages. It gives a characteristic flavor to the product, affects its external color after generation of furfural compounds from the reaction of smoke carbonyls with amino groups of proteins, especially when applied at high temperatures, contributes to preservation due to the bactericide and bacteriostatic effect of smoke compounds and exerts antioxidant properties due to the phenols present in the smoke. Some undesirable effects of smoking are associated with formaldehyde and other possible carcinogenic substances contained in process smoke (Bem *et al.* 1995).

Ripening/Drying

The main objectives of ripening/drying are to achieve the desired water activity and develop the characteristic flavor and appearance. The sausages hanging in the racks are placed in drying chambers (Fig. 4.7) where they are ripened at different time-temperature conditions depending on the size and type of product.

To achieve the desired water activity in the sausage it is necessary to control the drying of the sausage in a slow and steady way, which is a difficult task. There is a delicate equilibrium between two physical processes, water diffusion from inside the sausage to the outer surface and evaporation from the outer surface to the environment. Both processes, diffusion and evaporation, must proceed at similar rates to have the correct drying (Fig. 4.8). Some of them are related to the characteristics of the sausage itself, with pH and case diameter being the most important. But other factors, of great importance, are related to the chamber environment. So, if the relative humidity in the chamber is decreased too fast, there is a rapid drying that results in case hardening, an excessive dry and hard outer layer, slowed water diffusion and, consequently, evaporation.

(A)

(B)

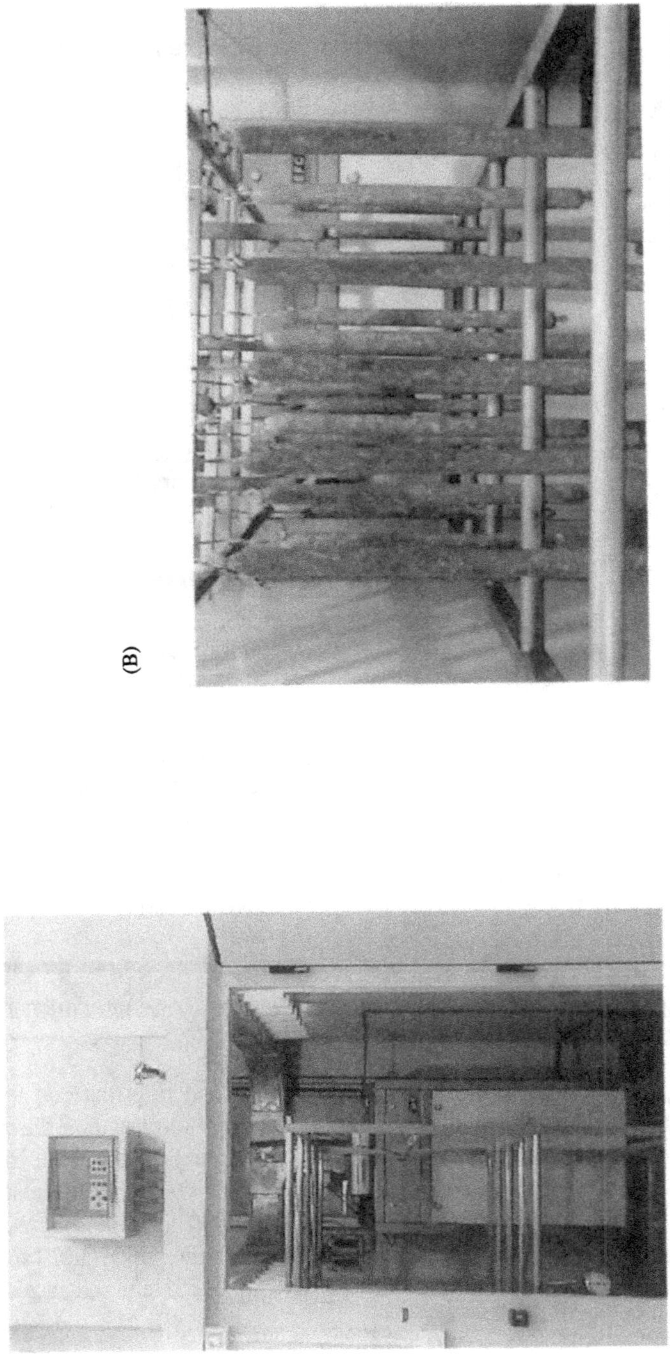

FIG. 4.7. (A) EXAMPLE OF A DRYING/RIPENING CHAMBER. (B) SPANISH "SALCHICHON" IN DRYING/RIPENING CHAMBER

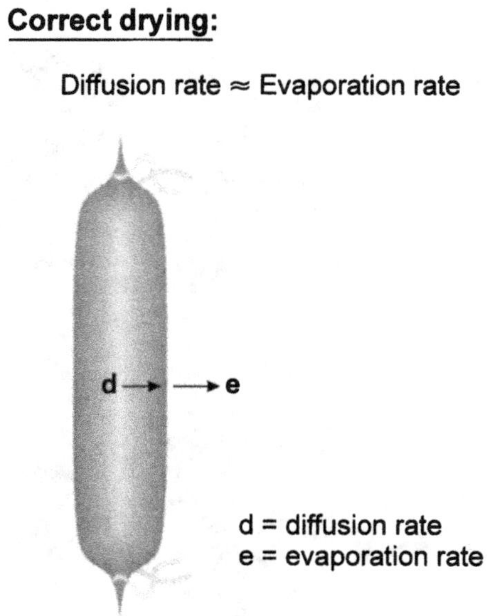

Main factors affecting drying rate

Inner sausages:	pH Amount of fat Particles size
Outer sausages:	Sausage diameter Pore diameter of casing Presence of molds on external surface
Chamber environment:	Temperature Relative Humidity Air rate

FIG. 4.8. SCHEME OF THE DRYING PROCESS IN A DRY-FERMENTED SAUSAGE

To solve this equilibrium, there is a practical rule that consists of keeping the relative humidity in the chamber about 5–10% lower than the water activity value (expressed as a percentage) of the sausage (Lücke 1985). In general, the temperature during this stage is mild, around 12–16C, and the relative humidity is decreased in a progressive way (RH from 90 to 65%). Some air velocity (around 0.05–0.1 m/s) is recommended to homogenize the humidity in the chamber environment. In general, the starting fat content and case diameter have increased sausage percent yields (Palumbo *et al.* 1976a). The casing must stay attached to the mixture when the sausage shrinks during drying. In addition to drying, there are important chemical and biochemical changes that are responsible for sensory properties.

Final Product

Dry-fermented sausages are usually sold as an entire piece that can be then cut into thin slices by consumers at home just before consumption. There is also a growing market for packages containing thin slices of ready-to-use sausage. These slices, once the casing is peeled-off, may be stored either in vacuum-packages or under controlled or modified atmosphere. The shelf life may be very long, higher than 9 weeks, with adequate manufacture, transportation and refrigerated storage.

CHANGES DURING THE PROCESSING

The main physical and biochemical changes that take place during the processing of dry-fermented sausages are listed in Fig. 4.9. The main biochemical changes, proteolysis and lipolysis, are described in Chap. 6 and 7, respectively. A description of the main physical changes are detailed below.

Color

The color of meat and meat products is influenced by its moisture and fat content and also by the content of hemoprotein, particularly myoglobin and its relationship with the environment surrounding it. The development of the characteristic color of fermented products is mainly due to the action of nitrite with myoglobin, which produces the red color (Pegg and Shahidi 1997). The lightness of the color depends on the zone of the sausage. The color is lighter in the center due to the higher content in moisture and darker in the outer surrounding area. The lightness value L, measured with the Hunter Lab system, increases during ripening but its values are lower in the periphery and higher in the central area caused by the moisture gradient into the sausage (Pérez-Alvarez *et al.* 1999). Most color determinations are made on homogenized samples instead of the intact slice. The reason is that direct measurement in the slice may produce variations of the path of the incident beam depending on the lean to fat ratio.

Nitrates and nitrites constitute the main curing agents added in the formulation of fermented sausage. The former is reduced to nitrite through the action of microbial nitrate reductase activity, typically present in Micrococceae. Most of the formed nitrite is reduced to nitric oxide either by nitrite reductase or by chemical reactions favored by ascorbic/erythorbic acid. The formation of curing color in fermented sausages is obtained through several steps (Lücke 1985). First, oxygenated myoglobin (red) reacts with nitrous acid to give metmyoglobin (brown) and nitrate. Second, nitrous acid is reduced to nitric oxide, favored by the presence of ascorbate/erythorbate, and metmyoglobin to myoglobin. Then, myoglobin and nitric oxide interact forming nitric oxide myoglobin (red). This reaction is favored at low pH and is thus accelerated by the activity of lactic acid bacteria. During ripening, the protein moiety of nitric oxide myoglobin is denatured giving the formation of nitric oxide myochromogen, which improves color stability. However, the peroxides from the fatty tissue or formed by lactic acid bacteria in the presence of oxygen, may

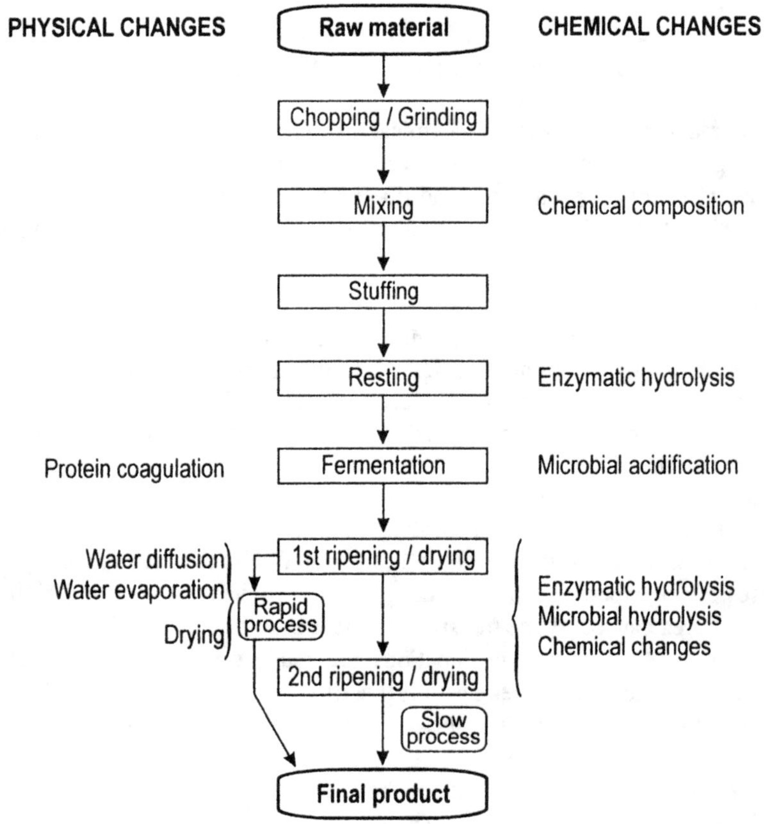

FIG. 4.9. MAJOR STEPS, WITH INCIDENCE OF MAIN PHYSICAL AND CHEMICAL CHANGES, IN THE MANUFACTURE OF A TYPICAL DRY-FERMENTED SAUSAGE

oxidize the iron within the porphyrin ring changing the color to gray or brown.Catalase is a microbial enzyme with an important role for color and flavor preservation of the product since it is able to protect both from oxidation by eliminating the peroxides (Demeyer 1992).

Texture

Texture development is the result of chemical and physical processes. The salt added in the formula gives a suitable cohesion and texture during drying by solubilizing proteins, which act as a bridge between the constituting meat fragments. Once the fermentation is initiated, there is an accumulation of acid substances resulting from the microbial activity which produces a progressive decrease in pH value. As pH approaches the isoelectrical point of meat proteins, water binding capacity is decreased, releasing free water. On the other hand, there is a continuous decrease in the solubility of myofibrillar proteins with a trend to coagulate and increase the product consistency. This consistency will also be accelerated during the drying process.

The consistency varies widely, from rigid for very dry sausages to soft when there is an excess of fat and water. It depends on the degree of drying, pH drop, amount of fat and extent of proteolysis, especially on myofibrillar proteins. The dissolved myofibrillar proteins are initially transformed into a thin fluid colloidal transition state with unstable coagulation bonds. Once proteins are denatured by lactic acid accumulation and loss of water progresses during drying, the bonds are stabilized and the matrix of the fermented sausage is developed (Katsaras and Budras 1991).

Connective tissue proteins also contribute to the structure although in a minor degree. The consistency may also vary within the sausage tending to be softer in the center and harder in the rind. During drying there is a continuous development of the textural characteristics of firmness, hardness and cohesiveness of meat particles. Shear values during drying are usually correlated with drying time, sausage moisture content, sausage diameter and sometimes depend on the initial grinding (Acton 1977). A good consistency is necessary to achieve the sliceability typical of the product. Firmness development is more related to the effects of pH on myofibrillar proteins. On the other hand, salt is more related to determine textural aspects associated with binding the meat mass together.

Humidity

The means to achieve the desired water loss and consequent decrease in a_w have changed in the last few decades. Today, most of the drying methods use air-conditioned computer-controlled plants. The final water loss is below 20% for semi-dry sausages and above 20% for fully dried sausages. In these cases, the final water content is usually found between 20 and 35%. The a_w levels are between 0.82 and 0.90. An example of humidity and a_w evolution is given in Fig. 4.10(A). The reduction of water activity is slower in those sausages containing beef because this meat is more resistant than pork in the process of desiccation. The rate of water loss depends on pH drop, which can be more or less pronounced depending on the accumulation of lactic acid. As pH reduces, meat proteins approach their isoelectric state. As the number of negative charges decreases, the water-binding capacity of proteins is also reduced and bound water is released. In general, when pH drops rapidly the sausage dries faster.

pH

The pH experiences an initial fall during the fermentation stage due to the accumulation of lactic acid produced by lactic acid bacteria. An example of the evolution of pH is given in Fig. 4.10(B). The rate of pH drop may vary depending on the amount and type of starter cultures added to the mix, amount and type (complex or simple) of sugars, time and temperature for fermentation, etc. The pH can rise during the ripening/drying due to the buffering action of proteins as well as the enzymatic formation (by peptidases and deamidases) of nonprotein nitrogen basic compounds and ammonium ions.

MAIN FERMENTED SAUSAGE VARIETIES

There are many varieties of sausages. The denominations vary according to the geographical area, more or less extensive depending on each product, where fermented sausages are produced and consumed. In some cases, and, for a given type of product, several additional names may be given depending on the form, caliber (diameter of cross section), size of fat particles, application of smoking, addition of specific spices, presence of external molds, etc. In the case of salamis, for instance, one can find different diameters and processing durations. Salamis having a 55–60 mm diameter, like French Ménage (ripened for 28–32 days), Italian Turista (ripened for 14–21 days) and Spanish Salchichón (ripened for 14–18 days), need relatively short processing times. However, salamis like French Varzi and Italian Crespone Milano with larger diameters, such as 90–100 mm, require ripening times as long as 60–70 days (Baldini *et al.* 2000).

Some of the typical Mediterranean sausages are French saucisson, Spanish chorizo or Italian salami (Fig. 4.11). On the other hand, German or Hungarian style salamis represent some of the typical North-European products. Pepperoni and summer sausage are highly popular in the U.S.

(A)

(B)

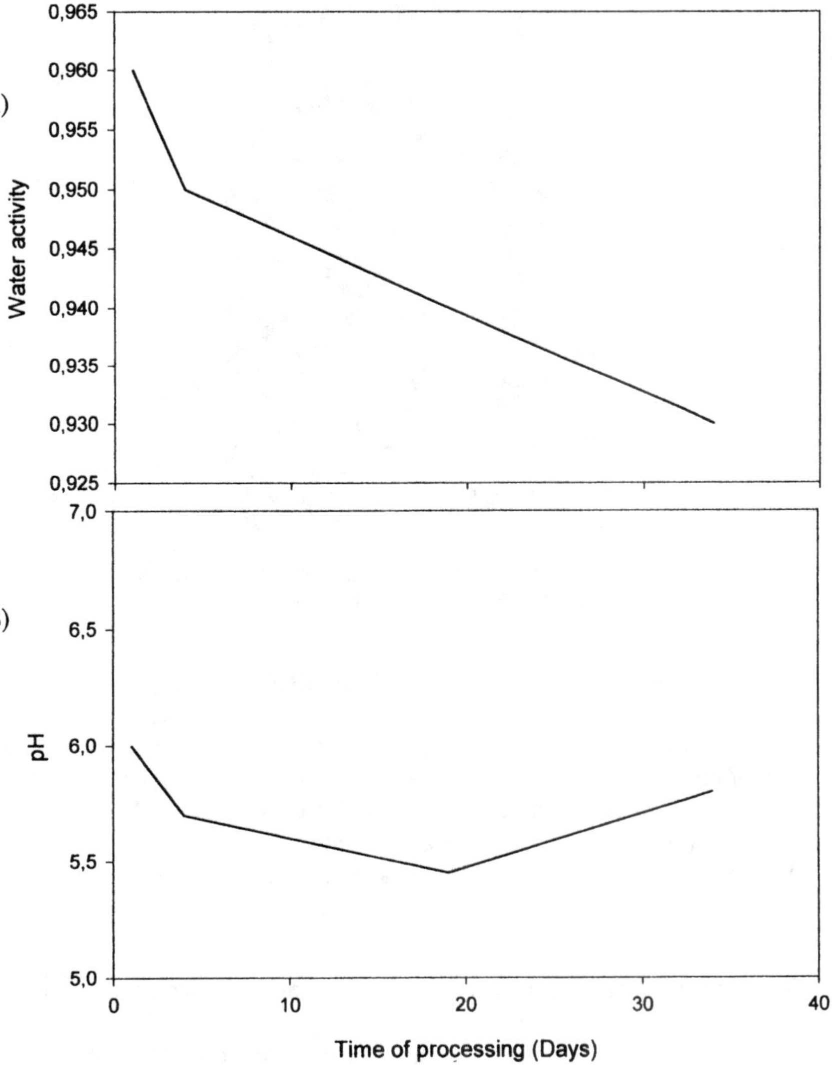

FIG. 4.10. EXAMPLE OF THE EVOLUTION OF a_w AND pH DURING THE PROCESSING OF A DRY-FERMENTED SAUSAGE

FIG. 4.11. ITALIAN SALAMIS DURING RIPENING/DRYING
(Courtesy of Mr. P. Baldini, Stazione Sperimentale di Parma, Italy)

TRENDS IN ACCELERATED PROCESSING

Many attempts to accelerate the ripening of dry-fermented sausages, reduce the time of processing and improve the flavor have been made. These attempts, which are attractive from the economical point of view, have been based on three major strategies: the use of enzymes, the use of new strains as starter cultures and the use of improved strains as new starter cultures.

Use of Enzymes

There are several advantages in the use of enzymes, such as the variety of enzyme activities available, their high specificity and their high catalytic rates under mild reaction conditions in the sausage mix. Most of the attempts made have been in the addition of commercial proteases and lipases from different sources, alone or in combination, to the sausage mix with the intention to enhance proteolysis and lipolysis, respectively, and accelerate flavor development. Most of these studies were originally developed for cheese and afterward applied to sausages based on the fact that the main concept of making cheese and sausages is very similar.

Different commercial proteases, such as pronase E from *Streptomyces griseus*, an aspartyl proteinase from *Aspergillus oryzae*, papain from *Carica papaya* and neutrase from *Bacillus subtilis,* have been assayed in Spanish-type dry-cured sausages (Melendo *et al.* 1996; Diaz *et al.* 1997; Zapelena *et al.* 1997). Even though an intense proteolysis was obtained, as detected by high amounts of released peptides, free amino acids and amines, the sensory quality was not significantly improved, but a lack in consistency and/or bitter taste were frequently reported.

In the case of Northern-type dry-fermented sausages, a substantial amount of shortened time for processing, up to 50%, was achieved by adding a serine proteinase from *Lactobacillus paracasei* subsp. *Paracasei* NCDO151, but the addition of alcalase, another commercial protease, did not improve the process (Hagen *et al.* 1996). In some cases, the excess of proteolytic activity results in texture problems, especially an excessive softening of the sausages. So, it is very important to select the correct dose of the proteinase for an adequate proteolysis and acceleration of the maturation without affecting texture.

The addition of a pancreatic lipase resulted in the generation of a high number of free fatty acids, sometimes even with an oily exudate appearance, but with no noticeable improvement in flavor (Ordoñez *et al.* 1999). Something similar was observed with the addition of a lipase from *Rhizomucor miehei* (Zalacain *et al.* 1997). It is evident that proteolysis and lipolysis provide great amounts of free amino acids, small peptides and free fatty acids that constitute flavor precursors, especially for the final aroma, but the addition of proteases and lipases alone is not enough. Further reactions to form volatile compounds are necessary for correct flavor development. This is a dynamic research area where new enzymes with improved or more accurate particular specificities may give better results.

The search for new enzymes involves selecting new microorganisms by traditional microbiological methods using enrichment cultures and selective media. However, it should be taken into account that the industrial production of microbial enzymes depends on its location in the cell. Extracellular enzymes are easier to produce and isolate, but the production may be costly and time-consuming if the enzymes are cell-bound and produced at low rates. This has prompted the search for new strains with desirable enzymatic activity or the genetic improvement of strains.

Use of Selected Starter Cultures

The inoculation of starter cultures has extended use and application worldwide for speed-up and control of the fermentation process, especially because of safety reasons derived from pH reduction. Traditionally, the starter industry selects those strains having the most desired characteristics and properties. Great efforts have been made in recent years for the selection of new strains from the meat environment. These microorganisms are usually chosen based on the following: presence along the curing process, predominance in the microbial flora, metabolic properties of interest for the final quality and absence of pathogenic activity. Then, those strains that have better metabolic activity or give higher yields of enzyme of technological interest within the shortest possible fermentation time are selected by classical methodologies and assayed as starter cultures. Another strategy relies on cell modification of bacterial cells to enhance flavor in shorter ripening times. So, desirable intracellular enzymes such as peptidases may be released to the sausage mix after lysis of the cell by physical methods (freezing, heating, chemical treatments or drying) and exert a faster action on their substrates.

Use of Improved Strains as New Starter Cultures

There is an increasing demand of strains with improved properties of interest for meat fermentation. The improvement of starter strains by using traditional methodologies has been done in the past, although it requires high efforts in personnel, cost and time. Modern biotechnology allows the genetic improvement of starter cultures in an easy and effective way. The DNA coding for desirable traits can be part of a plasmid or chromosome. The properties that reside on a plasmid can be transferred by conjugation that is generally regarded as safe (GRAS). Many research works are currently in progress. Lactobacilli, which lack pathogenicity and have many applications in foods, constitute an interesting group of microorganisms for genetic improvement. For instance, there is the overproduction of bacteriocins against *Listeria monocytogenes* (Leistner *et al.* 1991) or other genes (e.g., lysostaphin gene of *S. simulans*) for enhancing antimicrobial potential (Cavadini *et al.* 1996). Other examples include the suppressed production of mycotoxins and regulation of metabolic activity in molds (Leistner *et al.* 1991; Geisen 1993) and the transfer of genes (lipases) from other Gram-positive bacteria into *S. carnosus* (Goetz 1990; Al-Masaudi *et al.* 1991). However, the use of genetic modified organisms (GMOs) in foods is still controversial and is generally unaccepted by European consumers.

REFERENCES

ACTON, A.C. 1977. The chemistry of dry sausages. Proc. Annual Reciprocal Meat Conf. *30*, 49–62.

AL-MASAUDI, S.B., DAY, M.J. and RUSSELL, A.D. 1991. Antimicrobial resistance and gene transfer in *Staphylococcus aureus*. J. Appl. Bacteriol. *70*, 279–290.

BACUS, J. 1984. Update: Meat fermentation 1984. Food Technol. *38*, 58–63.

BEM, Z., ZLENDER, B. and HECHELMAN, H. 1995. Desirable and undesirable effects of smoking meat products. Die Fleischerei *3*, 3–8.

BLOM, H., HAGEB, B.F., PEDERSEN, B.O., HOLCK, A.L., AXELSSON, L. and NAES, H. 1996. Accelerated production of dry fermented sausage. Meat Sci. *43*, S229–S242.

BUEGE, D.R. and CASSENS, R.G. 1980. Manufacturing summer sausage. Cooperative Extension Programs, A3058, University of Wisconsin, Madison, WI.

CASSENS, R.G. 1995. Use of sodium nitrite in cured meats today. Food Technol. *49*, 72–81.

CASSENS, R.G. 1997. Composition and safety of cured meats in the USA. Food Chem. *59*, 561–566.

CAVADINI, C., HERTEL, C. and HAMMES, W.P. 1996. Stable expression of the lysostaphin gene in meat lactobacilli by introducing deletions within the prosequence. Syst. Appl. Microbiol. *19*, 21–27.

DEL MONTE, P., MAGNANI, U. and MONARI, M. 1990. Industria dei salumi. Edizioni Agricole, Bologna, Italy.

DIAZ, O., FERNÁNDEZ, M., GARCÍA DE FERNANDO, G., DE LA HOZ, L. and ORDÓÑEZ, J.A. 1997. Proteolysis in dry fermented sausages: the effect of selected exogenous proteases. Meat Sci. *46*, 115–128.

EUROPEAN DIRECTIVE 1995. Additives different from colorants and edulcorants used in foods. Ref. 95/2/CE of 20 February. European Commision. Brussels, Belgium.

FLORES, J. 1989. La charcuterie typiquement espagnole. Viandes Prod. Carnees *10*, 3–9.

FLORES, J. and BERMELL, S. 1996. Dry cured sausages. Factors influencing souring and their consequences. Fleiscwirtsch. *76*, 163–165.

FRENTZ, J-C. and ZERT, P. 1990. L´encyclopédie de la Charcuterie. 3rd Ed., pp. 624–626. Soussana, Paris, France.

GEISEN, R. 1993. Fungal starter cultures for fermented foods: molecular aspects. Trends Food Sci. Technol. *4*, 251–256.

GOETZ, F. 1990. Applied genetics in the Gram positive bacterium *Staphylococcus carnosus*. Food Biotechnol. *4*, 505–513.

GOKALP, H.Y. and OCKERMAN, H.W. 1985. Turkish-style fermented sausage (soudjouk) manufactured by adding different starter cultures and using different ripening temperatures. I. Fleischwirtsch. *65*, 1235–1240.

HAGEN, B.F., BERDAGUE, J.L., HOLCK, A.L., NAES, H. and BLOM, H. 1996. Bacterial proteinase reduces maturation time of dry fermented sausages. J. Food Sci. *61*, 1024–1029.

KATSARAS, K. and BUDRAS, K.-D. 1991. Microstructure of fermented sausage. Meat Sci. *30*, 121–134.

LEISTNER, L., GEISEN, R. and BOECKLE, B. 1991. Possibilities and limits to genetic change in starter cultures and protective cultures. Fleischwirtschaft. *71*, 682–683.

LÜCKE, F.K. 1985. Fermented sausages. *In* Microbiology of Fermented Foods. Vol. 2, B.J.B. Wood (ed.), pp. 41–83. Elsevier Applied Science, London.

MELENDO, J.A., BELTRAN, J.A., JAIME, I., SANCHO, R. and RONCALES, P. 1996. Limited proteolysis of myofibrillar proteins by bromelain decreases toughness of coarse dry sausage. Food Chem. *57*, 429–433.

NOVELLI, E., ZANARDI, E., CHIRETTI, G.P., CAMPANINI, G., DAZZI, G., MADARENA, G. and CHIZZOLINI, R. 1998. Lipid and cholesterol oxidation in frozen stored pork, salame Milano and mortadella. Meat Sci. *48*, 29–40.

ORDOÑEZ, J.A., HIERRO, E.M., BRUNA, J.M. and DE LA HOZ, L. 1999. Changes in the components of dry-fermented sausages during ripening. Crit. Rev. Food Sci. Nutr. *39*, 329–367.

PALUMBO, S.A., SMITH, J.L. and ZAIKA, L.L. 1976a. Sausage drying: Factors affecting the percent yield of pepperoni. J. Food Sci. *41*, 1270–1272.

PALUMBO, S.A., ZAIKA, L.L., KISSINGER, J.C. and SMITH, J.L. 1976b. Microbiology and technology of the Pepperoni process. J. Food Sci. *41*, 12–17.

PÉREZ-ÁLVAREZ, J.A., SAYAS-BARBERÁ, M.E., FERNÁNDEZ-LÓPEZ, J. and ARANDA-CATALÁ, V. 1999. Physicochemical characteristics of Spanish type dry-cured sausage. Food Res. Int. *32*, 599–607.

ROCA, M. and INCZE, K. 1990. Fermented sausages. Food Rev. Int. *6*, 91–118.

ZALACAÍN, I., ZAPELENA, M.J., PAZ DE PEÑA, M., ASTIASARÁN, I. and BELLO, J. 1997. Application of Lipozyme 10,000L (from Rhizomucor miehei) in dry fermented sausage technology: study in a pilot plant and at the industrial level. J. Agric. Food Chem. *45*, 1972–1976.

ZAPELENA, M.J., ZALACAÍN, I., DE PEÑA, M.P., ASTIASARÁN, I. and BELLO, J. 1997. Addition of a neutral proteinase from *Bacillus subtilis* (neutrase) together with a starter to a dry fermented sausage elaboration and its effect on the amino acid profiles and the flavor development. J. Agric. Food Chem. *45*, 472–475.

CHAPTER 5

FERMENTATION AND STARTER CULTURES

The use of starter cultures is currently practiced in the production of dry-fermented sausages and is described later in this chapter. The exceptions, which rely on the natural microflora, are used with traditional sausages fermented at low temperatures and over an extended period of time for ripening.

MICROBIOLOGY OF DRY-FERMENTED SAUSAGE

The origin and composition of the microflora naturally present in the raw sausage mix is diverse. There are many factors that determine which microflora are present, such as the hygiene in meat manipulation, the environment and microorganisms present in the additives. However, the presence of salt and nitrite, oxygen depletion, pH drop, reduction in a_w and accumulation of certain metabolites, like bacteriocins, during the processing will result in a kind of selectivity that favors the development of *Kocuria*, *Staphylococcus* and lactic acid bacteria but will prevent the growth of undesirable microflora like pathogenic and spoilage microorganisms (Leistner 1992). A typical microbial evolution along the fermentation/ripening of dry-fermented sausages is shown in Fig. 5.1(A).

Most of the microbial groups grow exponentially at the fermentation stage due to both temperature rise and pH decrease. Lactic acid bacteria with initial counts around 10^3–10^5 cfu/g may reach counts as high as 10^7–10^9 cfu/g during the fermentation stage since the genus Lactobacillus is the most competitive followed by Leuconostoc, Pediococcus and Streptococcus. *L. plantarum* is mostly found in sausages fermented at higher temperatures while the species *L. sakei* and *L. curvatus* dominate the flora at mild fermentation temperatures (20–24C), which are typical of traditional European fermented sausages (Toldrá *et al.* 2001). Heterofermentative lactobacilli usually represent less than 10% of lactic acid bacteria (Samelis *et al.* 1994; Kröckel 1995).

There is also a sensible growth of Micrococcaceae, although growth can be restricted depending on acid generation, from initial levels of about 10^3–10^5 cfu/g up to final counts around 10^6–10^7 cfu/g, especially *Staphylococcus*, which has the advantage to grow and metabolize in anaerobic conditions, and *Kocuria*.

The strains found in naturally fermented products mainly belong to the species *S. xylosus, S. carnosus* and *K. varians*. Yeasts and molds, initially present at low levels of 10^2–10^3 cfu/g or cm^2 (Roncalés *et al.* 1991), also increase up to 10^6–10^7 cfu/g or cm^2 during the fermentation stage. Yeasts, especially *Debaryomyces hansenii*, tend to grow toward the outer and external surface of sausages although they can be found in the center of the product at the start of the ripening (Leistner 1992). On the other hand, molds like *Penicillium* spp. generally grow only on the external surface.

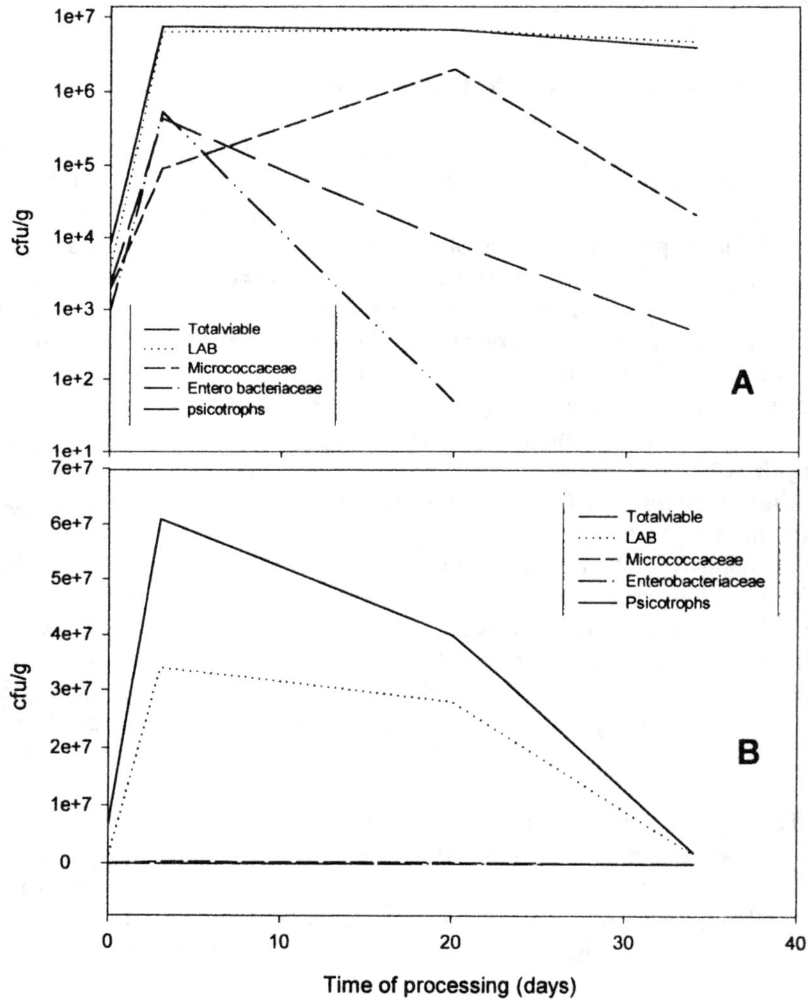

FIG. 5.1. TYPICAL MICROBIAL EVOLUTION ALONG THE PROCESSING OF A DRY-FERMENTED SAUSAGE (SPANISH SALCHICHÓN)
(A) Nonstarter added, (B) Addition of starter culture
(Adapted from Sanz *et al.* 1997)

There are several factors in the sausage that interact and act as consecutive barriers, defined as the "hurdle effect," against the growth of undesirable microorganisms (Leistner 1992). Thus, the inhibition of food-borne pathogenic and spoilage microorganisms is the result of the combined action of salt and nitrite addition, the amount and type of sugar added, the reduction in redox potential, the initial pH and a_w, the fermentation temperature, the initial number of lactic acid bacteria added as starter and the rate of acid formation. The growth of undesirable competitive microflora, due to the use of raw materials with poor quality or nonhygienic handling, constitutes the most important cause of faulty raw sausage production.

Enterobacteriaceae, which can cause core putrefaction, can increase slightly, up to 10^5 cfu/g, during the fermentation stage but rapidly decrease to almost negligible levels during the drying period. The growth of pathogenic bacteria such as *Salmonella spp.* is mainly prevented by the presence of nitrite in the initial stages and the further reduction in pH. Starter cultures containing lactic acid bacteria are inhibitory to the growth of *Salmonella spp*. This effect is more or less pronounced depending on the species and strain, the temperature and acid production (Roca and Incze 1990).

Listeria monocytogenes is a food-poisoning bacteria of special interest lately. This bacteria is inhibited by the low pH, competitive flora and accumulation of antimicrobial compounds. As bacteriocins from lactobacilli and/or pediococci have a good antagonistic effect against *L. monocytogenes*, the use of these lactic acid bacteria as starters would have an additional role (Hammes and Tichaczek 1994). For instance, the presence of a bacteriocin-forming pediococci inhibit the growth of *L. monocytogenes* during fermentation (Lücke 1992). Its count decreases with ripening when growth is not possible (Leistner 1992).

Staphylococcus aureus is a facultative anaerobe bacteria that is able to ferment a high number of diverse carbohydrates. It is salt tolerant, can grow at low a_w and is capable of producing the enterotoxin at fermentation temperatures but only under aerobic conditions (Roca and Incze 1990). These bacteria are sensitive to acids and are very poor competitors. The initial pH and the activity of the lactic acid bacteria are important for the control of *Staphylococcus aureus*. The risk is low in mild fermented sausages when initial counts are below 10^4 cfu/g. However, in the case of sausages fermented at high temperatures, it is necessary to control the process and get a rapid pH drop early in fermentation because *Staphylococcus aureus* could grow in the outer layers of the sausage during the lag phase of lactobacilli (Lücke 1985).

The presence of *Clostridium botulinum* and *C. perfringes* is excluded by the effect of nitrite combined with a decrease in pH and a_w (Lücke 1985). The growth of undesirable molds must be controlled in order to avoid mycotoxins. Smoking or immersion in potassium sorbate constitutes an effective measure in preventing the growth of undesirable molds (Leistner 1992).

METABOLISM

The microbial enzyme system comprises a good number of enzymes involved in different tasks. The activity of these enzymes depends on the final number of microorganisms or the lysis of the cells and the conditions in the sausage that change significantly during ripening. These conditions include pH, which is essential as each enzyme has an optimum pH, the NaCl content, since this compound is well known to exert activation or inhibition, ripening temperature, since enzyme activity increases with temperature and the water content, which affects composition and protein conformation.

The enzyme activity depends on the protein conformation of the substrate (in the case of proteases) and of the enzyme itself. The diffusion coefficients decrease with lower water content and affect reaction rates.

The microbial enzyme system is complex but may be classified as extracellular enzymes secreted to the media by the intact cell, cell-wall associated enzymes, cell-membrane associated enzymes and intracellular enzymes released after the lysis of the cell. All these enzymes can act during the fermentation and ripening of dry-fermented sausages and will have different roles depending on which type of reaction is catalyzed.

Sugar Metabolism

Sugar metabolism involves its transport into the cell and its further metabolism. Lactic acid is the main product resulting from carbohydrate fermentation. The generation and ratio of L and D lactic acid enantiomers in the final product will depend on the species of lactic acid bacteria present, and specifically on the action of L and D lactate dehydrogenase, respectively. The presence of lactate racemase may also affect the ratio of the enantiomers in the racemic mixture.

Once the carbohydrate has been transported into the cell, sugar metabolization occurs via the glycolytic or Embden-Meyerhof pathway. The homofermentative lactic acid fermentation involves consecutive reaction steps catalyzed by several enzymes. Several important compounds are sequentially formed through the action of key enzymes: glyceraldehyde-3-phosphate by aldolases with the generation of NADH, pyruvate (the central intermediate in fermentation) from phosphoethanol pyruvate by pyruvate kinase, and lactic acid from pyruvate by lactate dehydrogenase with the oxidation of the NADH originated during the hydrolysis of glyceraldehyde 3-phosphate.

Most of glucose is decomposed in a homofermentative way. However, trace amounts of other end products such as acetate, formate, ethanol, acetoin, etc., may appear from the heterofermentative pathway. The quantity of sugar needed will depend on the type of sugar added, the curing agents present in the mixture and the process followed. The type of carbohydrate must be carefully chosen, since it affects the rate of pH drop. It must be chosen based on the temperature of fermentation, the ability of the strain to ferment it (Table 5.1) and the total time of processing. The amount of carbohydrate added will affect the extent of pH drop. Approximately 1% sugar will yield a reduction of about 1 pH unit during fermentation. The rate and extent of pH decrease, as a consequence of lactic acid accumulation, is very important for preventing the growth of undesirable microorganisms. As pH decreases and approaches the isoelectric point of most of the meat proteins, they coagulate and the consistency of the product increases. The redox potential is reduced during the lactic acid fermentation, keeping the anaerobic environment inside the sausage.

Proteolysis

An important hydrolysis of myofibrillar and sarcoplasmic proteins takes place during sausage fermentation and ripening. The myofibrillar structure is degraded and a significant amount of small peptides and free amino acids accumulate, contributing to flavor or indirectly as precursors of flavor. Proteolysis in dry-fermented sausages is brought about by the combined action of starter proteases, muscle proteinases (cathepsins and calpains) and exopeptidases, although the relative role of each group of enzymes is not clearly established yet. Endopeptidases would be mainly involved during fermentation whereas exopeptidases would play a major role during ripening (Demeyer 1992; Toldrá and Verplaetse 1995). There is a wide variety of proteolytic enzymes due to the high number of microorganisms with different enzymatic profiles used as starter cultures (Table 5.2), but, in general, these enzymes can be grouped into three major classes:

1) Proteinases (endopeptidases), responsible for the breakdown of protein into large polypeptides (most of them are cell-wall bound and are not secreted in the medium)

2) Peptidases, which are largely intracellular and are responsible for the generation of small peptides from polypeptides

3) Exopeptidases (also located intracellularly), responsible for the generation of free amino acids

The major steps of microbial proteolysis are shown in Fig. 5.2. Oligopeptides and small peptides are generated by the action of proteinase and peptidases located at the outer surface of the cell. These peptides are transported into the cell through transport systems. Once into the cell the peptides are further hydrolyzed into free amino acids by intracellular tri- and dipeptidases and aminopeptidases (Pritchard and Coolbear 1993). Proteolytic enzymes and transport systems from dairy starters like *Lactococcus* have been deeply studied and are used as a reference for the study of other microorganisms like *Lactobacillus* (Bockelman 1995). In the case of lactobacilli, several proteases located in the cytoplasm have been purified and characterized during the last decade. A major aminopeptidase from *L. sakei* with optimal pH 7.5 and temperature at 37C has been purified and characterized (Sanz and Toldrá 1997). This enzyme has a preference for the hydrolysis of leucine and alanine.

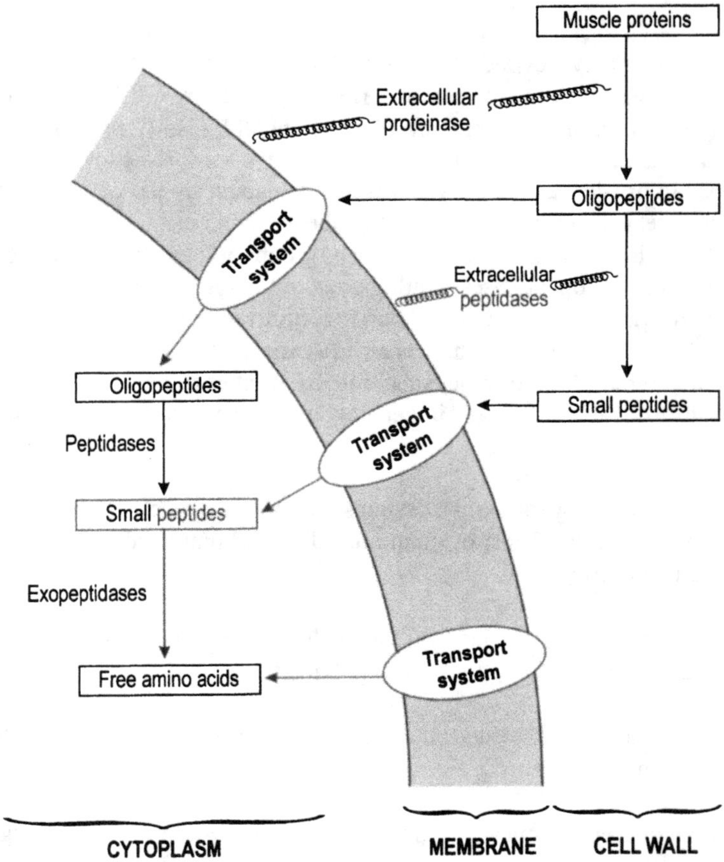

FIG. 5.2. SCHEME FOR THE POSSIBLE PATHWAYS IN THE HYDROLYSIS OF MUSCLE PROTEINS BY LACTIC ACID BACTERIA IN MEAT FERMENTATION

TABLE 5.1.
CARBOHYDRATE FERMENTATION CONDITIONS OF STARTER CULTURES USED FOR MEAT FERMENTATION

Microorganism	Species	Growth T (°C)	Acetoin Prod.	H_2O_2 Prod.	Lactic acid	Glucose	Saccharose	Lactose	Maltose	Starch
Lactobacillus	*sakei*	20-25	+	+	DL	+	+	+	-	-
" "	*curvatus*	"	±	+	DL	+	±	±	±	-
" "	*plantarum*	30-35	+	±	DL	+	+	+	+	-
" "	*pentosus*	"	+	±	DL	+	+	+	+	-
Pediococcus	*pentosaceus*	"	+	-	DL	+	+	+	+	-
" "	*acidilactici*	>40	+	+	DL	+	+	-	-	-
Kocuria	*varians*	30-35	-	-	DL	+	-	±	-	-
Staphylococcus	*xylosus*	"	+	-	DL	+	+	+	+	-
" "	*carnosus*	"	+	-	DL	+	-	±	-	-

+ property present; ± property depending on strain, - property absent
(Adapted from Lücke and Hechelmann 1987, and Hammes and Knauf 1993)

TABLE 5.2.
IMPORTANT ENZYMATIC ACTIVITIES OF STARTER CULTURES USED FOR MEAT FERMENTATION

	Microorganism	True catalase	Pseudo catalase	Nitrate Reducta activity	Nitrite Reduct Heme-depend.	Nitrite Reduct Heme-indep.	Endo protease activity	Exo Protease activity	Lipase activity	Deamidase activity
Bacteria	*L. sakei*	+	-	-	-	±	+	+	-	-
	L. curvatus	-	-	-	-	-	+	+	-	-
	L. plantarum	±	±	±	±	±	+	+	-	-
	L. pentosus	+	-	±	+	-	+	+	-	-
	P. pentosaceus	-	+	-	±	-	-	+	+	-
	P. acidilactici	+	-	-	-	-	-	+	-	-
	K. varians	+	-	+	-	+	-	+	+	-
	S. xylosus	+	-	+	-	-	-	+	+	-
	S. carnosus	+	-	+	-	-	-	+	±	-
Yeasts	*D. hansenii*	+	-	-	-	-	+	+	+	+
	S. specialis	+	-	-	-	-	+	+	+	+
	C. famata	+	-	-	-	-	+	+	+	+
Molds	*P. nalgiovense*	-	-	-	-	-	+	+	+	+
	P. chrysogenum	-	-	-	-	-	+	+	+	+

+ property present; ± property depending on strain, - property absent
(Adapted from Hammes *et al.* 1990, Lücke and Hechelmann 1987)

Other minor aminopeptidases, with acid and basic optimal pH, have also been purified and characterized. A tripeptidase from the same microorganism has an optimal pH 7.0 and optimal temperature at 40C and hydrolyzes different tripeptides at different rates depending on the specific sequence (Sanz *et al.* 1998). Recently, a dipeptidyl peptidase, which generates dipeptides from the N-terminal side of proteins and polypeptides, has been purified and characterized from *L. sakei* (Sanz and Toldrá 2001). A dipeptidase, which hydrolyzes dipeptides into two single free amino acids, has also been identified in *L. sakei* with an optimal pH at 7.6–8.0 and a temperature at 40–45C (Montel *et al.* 1995).

The proteinase and aminopeptidase activity of strains of *L. sakei*, *L. plantarum*, *L. casei* and *L. curvatus* assayed against synthetic substrates are shown in Table 5.3. These strains showed a wide specificity against amino acids at the N-terminal position. This was especially so for alanine, valine and leucine. When they were incubated with myofibrillar and sarcoplasmic proteins, their behavior differed. Therefore, the proteolysis was more intense on sarcoplasmic proteins, although peptides were generated for both groups of proteins. The highest amounts of free amino acids, as the final products of proteolysis, were released from sarcoplasmic proteins by *L. sakei* and *L. plantarum* and, in a minor scale, by *L. casei* and *L. curvatus*. However, fewer free amino acids were released when these strains were incubated with myofibrillar proteins (Fadda *et al.* 1999a, b; Sanz *et al.* 1999a, b).

Amino Acid Metabolism

Free amino acids, generated as final products of the proteolysis of muscle proteins, act as substrates for several enzymatic reactions like dehydrogenation, decarboxylation, deamination and transamination, which produce a wide range of compounds as schematized in Fig. 5.3. The main source of enzymes for most of these reactions are the microorganisms present in the product. Little is known about the enzymes involved and their regulation.

Dehydrogenation. Microbial NAD-dependent dehydrogenases transform the amino acids in the corresponding keto acid together with the generation of NH_4^+, producing a rise in the pH of the product. Glutamate dehydrogenase and alanine dehydrogenase generate α-ketoglutarate and pyruvate, respectively, and ammonia in the presence of NAD^+ or $NADP^+$. An example of the type of reactions carried out by glutamate dehydrogenase is:

$$\text{L-glutamate} + NAD^+ + H_2O \Leftrightarrow \alpha\text{-ketoglutarate} + NADH + NH_4^+$$

TABLE 5.3.
AMINOPEPTIDASE ACTIVITY (EXPRESSED AS A % OF ACTIVITY AGAINST ALA) OF STRAINS OF *L. SAKEI, L. PLANTARUM, L. CASEI* AND *L. CURVATUS* ON SYNTHETIC AMINOMETHYLCOUMARIN (AMC) SUBSTRATES

Substrate	*L. sakei*	*L. plantarum*	*L. casei*	*L. curvatus*
Alanine	100	100	100	100
Lysine	0.6	30.7	11.3	7.3
Serine	1.0	0.9	0.8	2.0
Phenylalanine	1.0	5.2	5.9	4.5
Valine	29.5	6.2	42.8	12.4
Arginine	2.4	13.7	15.0	13.3
Glycine	0.6	0.3	1.0	1.3
Leucine	48.9	20.5	94.9	78.8
Tyrosine	0.4	7.8	3.3	1.0
Proline	1.8	0.3	0.7	6.4

(Adapted from Fadda *et al.* 1999a, b, and Sanz *et al.* 1999a, b)

Decarboxylation. The microbial decarboxylation of certain amino acids leads to the formation of biogenic amines. The decarboxylation is mediated by enzymes from different microorganisms such as bacteria, molds and yeasts through the following reaction:

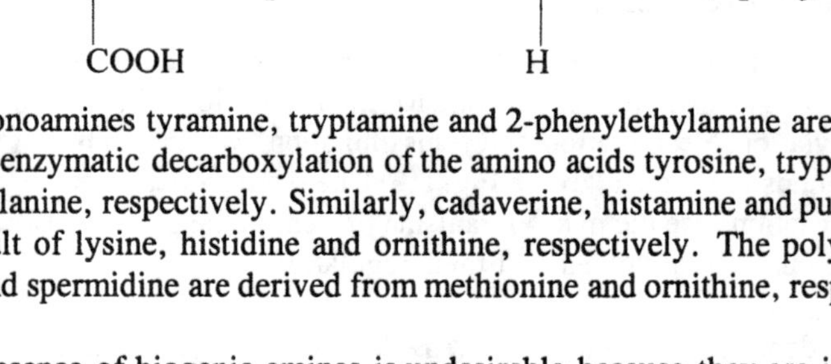

The monoamines tyramine, tryptamine and 2-phenylethylamine are formed through the enzymatic decarboxylation of the amino acids tyrosine, tryptophane and phenylalanine, respectively. Similarly, cadaverine, histamine and putrescine are the result of lysine, histidine and ornithine, respectively. The polyamines spermine and spermidine are derived from methionine and ornithine, respectively.

The presence of biogenic amines is undesirable because they are involved in nitrosamine formation and hypersensitive reactions in certain sensitive people (Monnet *et al.* 1996). The final concentration may vary depending on processing conditions and the bacterial decarboxylase activity, but, in any case, the maximum levels of total amines generated in dry-fermented sausages does not exceed 300 μg/g (Demeyer *et al.* 2000).

The total content of amines increases at a maximal rate during the fermentation stage (Dierick *et al.* 1974). It is important to control the growth of bacteria having decarboxylase activity and reduce its action to negligible levels in order to avoid the formation of biogenic amines. Good hygiene and

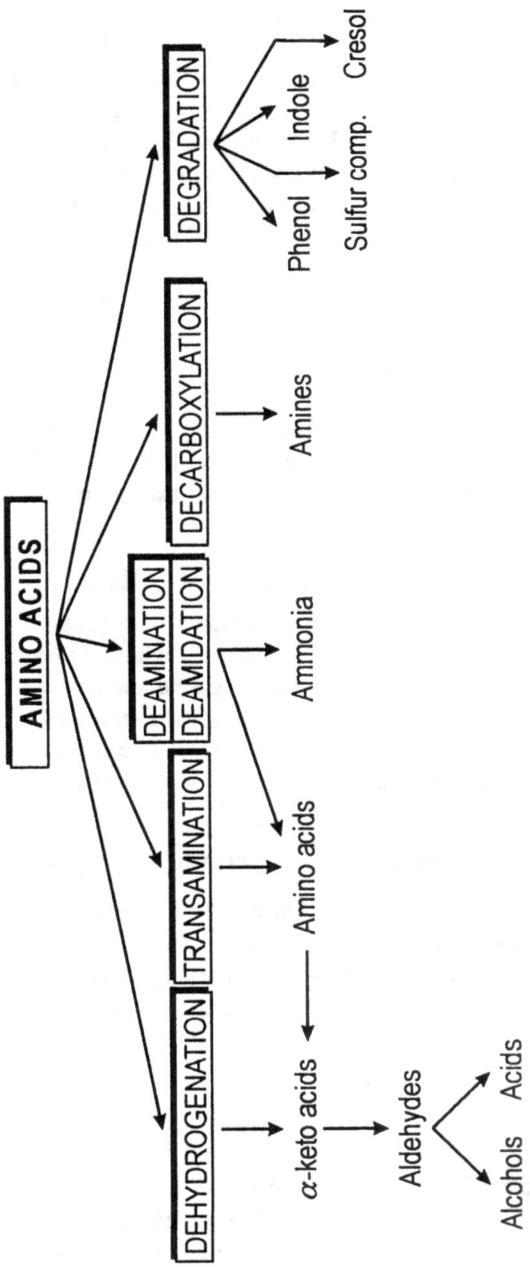

FIG. 5.3. MAIN REACTIONS INVOLVED IN AMINO ACIDS METABOLISM

appropriate raw materials, as well as correct processing temperature and use of lactic acid bacteria with the absence of decarboxylase activity, limit the level of biogenic amines in the products (Ordoñez *et al.* 1999).

Deamidation. This is a reaction of carboxylic acid amide hydrolysis and is facilitated by the presence of substitution in the β carbon atom of the amino acid. Asparaginase (L-asparagine aminohydrolase, EC 3.5.1.1.) and glutaminase (L-glutamine aminohydrolase, EC 3.5.1.2.) are enzymes usually present in yeasts and molds. These enzymes are capable of generating ammonia and producing a pH rise in the media. The main reactions catalyzed by asparaginase and glutaminase are, respectively:

$$\text{L-asparagine} + H_2O \Leftrightarrow \text{L-aspartate} + NH_3$$
$$\text{L-glutamine} + H_2O \Leftrightarrow \text{L-glutamate} + NH_3$$

Deamination. Deaminases catalyze the hydrolysis of different amino acids, especially aspartic and glutamic acids, but also others like threonine, serine and arginine. The result of these reactions is the removal of the amino group and the generation of ammonia.

Transamination. Transaminases or amino transferases are present in large amounts in bacteria. The reaction involves the transfer of the α-amino group of the first amino acid to the α-carbon atom from an α-ketoglutarate leaving behind the corresponding α-keto acid, from the first amino acid and a new amino acid. There is no loss of amino groups because the α-keto acid becomes aminated as the amino acid is deaminated (Lehninger *et al.* 1993). Transaminases differ in their substrate specificity. An example of the reaction catalyzed by alanine amino transferase is:

$$\text{L-alanine} + \alpha\text{-ketoglutarate} \Leftrightarrow \text{L-glutamate} + \text{pyruvate}$$

Degradation. Several microbial enzymes are able to decompose amino acids converting them into other compounds, most of them with specific flavors. This is the case of lyase which decomposes tyrosine into phenol or tryptophan into indol, ammonia and pyruvic acid. In other cases, several sulfur compounds may result from the action of demethiolase on sulfur-containing amino acids like methionine.

Lipolysis

Lipolysis constitutes a biochemical process with an important contribution to flavor development because free fatty acids with unsaturations will act as substrates for further oxidation to form volatile compounds with aroma properties (as discussed in Chap. 8 and 9, respectively). Traditionally, microbial species were selected as starter cultures according to their lipolytic activity for guaranteeing adequate flavor development in the sausage. But recent reports suggest there is an increased importance of endogenous muscle and fat lipases in relation to microbial lipases (Molly *et al.* 1997; Demeyer *et al.* 2000).

Fat tissue, which constitutes the major fat fraction of the sausage, is mainly composed of triglycerides. Intramuscular fat, although present in minor amounts of the sausage, also contains phospholipids, which are rich in polyunsaturated fatty acids.

Lipases, which are triacylglycerol acylhydrolase (EC 3.1.1.3.), act best at lipid-water interfaces and in any condition where an increase in the surface area of the substrate-water mixture would be expected to increase the activity of the enzyme.

Microbial lipases do not show stereospecificity in the hydrolysis of triglycerides. However, they may have different specificities like the hydrolysis of tri-, di- and monoglycerides at different rates, preference for types of fatty acids or positional specificity where the enzyme hydrolyzes the esters at preferential positions (Wong 1995). For instance, it has been observed in sausages that lipases prefer to hydrolyze the external positions of the triglyceride molecule (Ordóñez *et al.* 1999). Regarding the type of fatty acid, linoleic acid has been found to be the fatty acid most released, followed by oleic, stearic and palmitic acids (Demeyer *et al.* 1974). The major steps in the lipolysis of fat tissue are as follows:

$$\begin{array}{c} \text{Triglyceride} \rightarrow \text{Diglyceride} + \text{free fatty acid} \\ \downarrow \\ \text{Monoglyceride} + \text{free fatty acid} \\ \downarrow \\ \text{Glycerol} + \text{free fatty acid} \end{array}$$

Most of the lactic acid bacteria (LAB) cannot hydrolyze triglycerides, but they have mono- and diglyceride lipases that can act on mono- and diglycerides, respectively, and contribute to the generation of free fatty acids, especially those with short chains (Sanz *et al.* 1988). On the other hand, *Kocuria* has varied extra and intracellular lipolytic activity that can generate long- and short-chain free fatty acids such as acetic, propionic, isobutyric and isovaleric acids. The extracellular enzymes are most important because they remain in the medium, once the micrococci counts decrease, and can continue to hydrolyze fatty acids. The lipolytic activity of staphylococci, especially of *S. warneri* and *S. saprophyticus*, is also important, because it is lower in *S. carnosus* and *S. xylosus*, which only exhibit esterase activity (Montel *et al.* 1993). Yeasts and molds have strong lipase and esterase activity that can generate free fatty acids in the fat tissue at the outer surface of the sausage.

Nitrate Reductase

Nitrate reductase activity is very important for transforming the nitrate present in the sausage to nitrite. The main role of nitrite is to preserve the sausage against the growth of undesirable microorganisms (Cassens 1997). But nitrite is beneficial in many additional functions, such as the development and stabilization of typical cured-meat color (Flores and Toldrá 1993) and the prevention of lipid oxidation by binding heme, nonheme iron and stabilizing olefinic lipids against oxidation (Igene *et al.* 1985). The generated nitrite can be reduced to nitric oxide either by nitrite reductases or by further chemical reactions favored by ascorbic acid, the cysteine/cystine redox system or other redox systems (Ordoñez *et al.* 1999). *Kocuria* and *Staphylococcus* strains exhibit significant nitrate reductase activity. All staphylococci, except *S. warneri*, synthesize nitrate reductase with a maximal rate during the exponential growth phase. The production of nitrate reductase is increased in the presence of nitrate (Montel *et al.* 1999). In general, lactic acid bacteria lack nitrate reductase activity, although some strains of *L. plantarum*, *L. pentosus* and *P. Pentosaceus* exert nitrite reductase activity. This activity may be either heme-dependent or independent, although there are no limitations in activity because meat contains enough heme (Hammes *et al.* 1990).

Catalase

Catalase activity stabilizes color and flavor by its degradation of hydrogen peroxide. *Kocuria* and *Staphylococcus* strains are mainly involved in peroxide reduction through the true catalase or heme-dependent catalase that has no limitations in activity. The production of catalase in staphylococci is maximal at the end of the exponential growth phase and is the synthesis and release favored by the addition of nitrate (Montel *et al.* 1999). Lactic acid bacteria exhibits a second enzyme activity known as pseudocatalase or manganse-dependent catalase, but their contribution to the reduction in peroxides is minimal (Hammes and Knauf 1994). On the other hand, lactic acid bacteria like *L. sakei*, *L. curvatus* or *P. pentosaceus* can form peroxides, and some accumulation may be expected, depending on the catalase activity of the microbial population.

STARTER CULTURES

The use of starter cultures is a relatively new practice that has been progressively replacing traditional practices, which involve the use of indigenous flora or the inoculation of small portions of a previous fermentation mixture. The strains were selected by screening bacteria that are naturally found in sausages. Early attempts with Lactobacilli failed because of a lack of survival after lyophilization. *Pediococcus acidilactici* was firstly commercialized as a starter culture in the U.S. in 1957 (Everson *et al.* 1970), and *Pediococcus cerevisiae* was suitable for American summer sausages usually fermented at 30C, because of its easy lyophilized or frozen storage.

A different trend was followed in Europe, with low fermentation temperatures and nitrate addition. Europe faced the problem of reduced or even suppressed nitrate reductase activity due to a rapid acidification that resulted in frequent defects in color and flavor. A single strain of *Kocuria*, named M53, was the first one available in Europe (Niinivaara *et al.* 1964).

The next stage in the use of starter cultures involved the development of mixed cultures from a combination of strains with different metabolic properties. These strains were selected for their cooperative qualities and because they would not inhibit each other's activity. Because starters have a decisive effect on the quality of the final sausage, it is important to select those strains that have the most suitable properties, are in the most adequate proportions and keep the composition (Lücke 1985). Today, there are three main types of starter cultures. A single-strain culture is a pure culture made of one single strain. Multiple-strain cultures use either several different types of bacteria or different strains of one type. Lastly, mixed strains contain a mixture of strains of different types of bacteria.

Starter cultures were developed in response to safety improvements, an increase in industrial production, shorter processing times, the standardization of processes and products and the extended shelf life of products. The application of starter cultures is a common industrial practice. An example of microbial evolution is shown in Fig. 5.1(B). It is important to follow the manufacturer's instructions on storage and handling of starter cultures to be sure the microorganisms remain active. The current commercial starter cultures usually contain lactic acid bacteria (*Lactobacillus* and *Pediococcus*) to ensure a correct pH drop, and, in most cases, strains of *Kocuria* and *Staphylococcus* are added for additional lipolytic, catalase and nitrate reductase activity. The application of yeasts or molds is restricted to the external surface. The presence of molds on the external surface of the sausages contributes to the characteristic appearance and quality that are desirable in some Mediterranean areas.

There are three ways to develop a large and desirable microbial population in a sausage mix. First, there is the traditional process, in which the appropriate bacteria are introduced into the sausage mix by chance contamination from the meat, ingredients, additives, etc. The development of lactic acid bacteria during the cold storage will later contribute to lactic acid production, although in variable amounts, depending on the bacterial population. This variability in lactic acid, and consequently in pH drop, constitutes a high risk. Therefore, low temperatures are kept for a longer ripening period in order to avoid the growth and toxin production of harmful bacteria like *Staphylococcus aureus*.

Second is back inoculation or back slopping. In this method, 5 to 10% of the batter from an earlier good-quality batch, containing large numbers of lactic acid bacteria, is removed and added to the fresh sausage mix. There is a high risk of contamination even though a successful propagation of desirable bacteria is obtained.

Lastly, there are starter cultures. This modern practice involves the addition of high doses of lactic acid bacteria to ensure adequate acid production during fermentation and suitable quality during ripening.

Microbiology of Starter Cultures

Lactic Acid Bacteria (LAB). These bacteria belong to any of the following genera: *Lactobacillus*, *Leuconostoc*, *Enterococcus* and *Pediococcus*. The different commercial starter cultures are based on lactobacilli and pediococci and have different optimum temperatures for growth. For instance, *L. plantarum* and *Pediococcus* spp. at 30–35C and *P. acidilactici* at 40C are favored in fermentations at high temperatures and are typically used in the U.S. The major goal for the use of lactic acid bacteria is the pH drop of the meat mixture as a result of the acidification from carbohydrate metabolism. The pH drop and acidification are necessary to inhibit unwanted growth and toxin production of undesirable pathogens such as *Staphylococcus aureus*, *Salmonella* spp. and *Clostridium botulinum*. In the case of traditional European fermentation, where milder temperatures (20–24C) are used, the addition of *L. sakei* or *L. curvatus*, with optimal growth at 25–30C, is more convenient. In these cases, it is interesting to get a slower reduction in pH for allowing Micrococci to act and contribute to color and flavor.

Thus, the addition of lactic acid bacteria cultures in sausage processing has the following desirable effects: better safety by the reduction in pH, flavor due to characteristic acid taste and volatile compounds with aroma properties, coagulation of meat proteins by pushing pH to their isoelectric point (at pH $<$ 5.4–5.5) with consequences in texture and firmness and development of a desirable red color by favoring the reaction of nitric oxide with myoglobin giving nitrosomyoglobin.

Lactobacilli utilize glucose via the Embden-Meyerhof pathway (glycolysis), generating lactic acid as the major fermentation product. These organisms are all facultative heterofermentative. Therefore, undesirable fermentation products like acetic acid, hydrogen peroxide, carbon dioxide, acetoin and formic acid may be generated through the activation of the heterofermentative pathway. For instance, some strains may generate formate, acetate and small amounts of ethanol under anaerobiosis and conditions of glucose depletion. *L. sakei* and *L. curvatus* use oxygen to generate hydrogen peroxide and pyruvate, while *L. plantarum* oxidizes lactic acid into acetate and carbon dioxide. Although

Pediococcus are homofermentative organisms, *P. pentosaceus* may also produce acetate and ethanol from hexoses and pentoses (Kröckel 1995).

The enzymatic profile of lactic acid bacteria is shown in Table 5.2. Their proteolytic system is involved in muscle protein degradation with the subsequent release of small peptides and free amino acids and a partial contribution to flavor (Toldrá and Verplaetse 1995; Molly *et al.* 1997). In this sense, several studies carried out with *L. sakei*, *L. plantarum*, *L. carnosus* and *L. casei* have shown endo- and exopeptidase activity against muscle sarcoplasmic and myofibrillar proteins. Lactobacilli have an important exopeptidase system showing a good potential role in peptide degradation in meat fermentation.

The metabolic activity of some lactic acid bacteria may promote the formation of biogenic amines mainly by decarboxylation of amino acids. For instance, *L. curvatus* has shown decarboxylase activity that could produce up to four different amines (Hammes and Knauf 1994; Straub *et al.* 1995). Thus, it is important to select strains, like those from *L. sakei*, without the ability to decarboxylate amino acids. The competitive elimination of amine-producing strains is another strategy.

The production of bacteriocins, antimicrobial compounds of peptidic nature, by *Lactobacillus* is of great interest due to their application as spoilage prevention against undesirable organisms in meat products. Bacteriocinogenic strains of *L. sakei, L. curvatus, L. plantarum* and *Pediococcus* have been found, and several bacteriocins, like sakacin K from *L. sakei* CTC494, have been characterized, purified and their peptides sequenced (Hugas and Monfort 1997). Most research has been focused on *Pediococcus acidilactici,* a starter culture commonly used in American fermented meats. Some bacteriocins have shown good inhibition of *Listeria monocytogenes*. However, the effectiveness of a bacteriocin-producing culture in laboratory media does not guarantee its effectiveness in fermented sausages due to the complexity of the media (pH, curing agents, temperature, a_w and microbial growth) and the activity of endogenous muscle proteases.

Micrococcaceae. The most important strains commercialized as starter cultures are *Kocuria* and *Staphylococcus*. Their carbohydrate fermentation conditions and enzymatic profiles are shown in Table 5.1 and 5.2, respectively. Their nitrate/nitrite reductase activity is very important for the generation of nitric oxide and for color formation. In addition, they also stabilize the color and flavor by degrading the hydrogen peroxide with their catalase activity. These microorganisms are especially important when nitrates are added, but their sensitivity to low pH restricts their use to low-carbohydrate addition and long-term ripening sausages. Fermentation must progress slowly to allow the growth of Micrococci. Nitrate reductases transform the nitrate present in the sausage to nitrite. The nitrite generated from nitrate can be reduced to nitric oxide either by nitrite reductases or by further chemical reactions favored by ascorbic acid, the cysteine/cystine redox system or other redox systems (Ordoñez *et al.* 1999). Since *Micrococcaceae* grow little during the ripening process, it is necessary to add enough microorganisms to the sausage mix.

Most of these microorganisms have a desirable lipolytic activity, which is essential for the final flavor of the products. The lipase and esterase activity is involved in the release of free fatty acids and in aroma generation (Johansson *et al.* 1993; Stahnke 1995). The proteolytic activity of *Micrococcaceae* is generally low or even negligible, but some aminopeptidase activity has been detected in *K. varians*, *S. xylosus* and *S. carnosus* (Fransen *et al.* 1997).

Yeasts. The predominant species in fermented meats is *Debaryomyces hansenii*. Yeasts have an aerobic metabolism and grow preferentially on the surface of sausages, although some growth can be detected in the inner part of sausages. The application of selected yeast strains mainly contribute to the consumption of organic acids, especially lactic acid, and increase the pH through their deamidase activity. For these reasons, yeasts indirectly contribute to flavor by reducing the sour taste. In addition, they have some catalase activity, which delays the onset of rancidity, and proteolytic and lipolytic activity, which contributes to the final flavor.

Molds. *Penicillium nalgiovense* and *P. chrysogenum* constitute some of the most representative fungi available as starter cultures. They are inoculated on the surface of mold-fermented sausages and contribute to the appearance, flavor and safety of the final product. Molds have aerobic metabolism and their growth is restricted to the surface, giving a characteristic external appearance and contributing to flavor through the activity of lipases, proteases and deaminases. A thick layer of desirable molds on the surface serves as protection against the adverse effects of oxygen and light and prevents mycotoxigenic molds from becoming established.

Requirements for Starter Cultures

Strains used as starter cultures must be generally regarded as safe (GRAS) since they are considered food additives. There are some requirements that starter cultures must meet for effective use. They must be nonpathogenic, nontoxic or nonallergenic and must have genetic stability. Starter cultures also need to be tolerant of salt and nitrite; able to grow at manufacturing pH and temperatures; contribute to safety, flavor or nutrition; have nondecarboxylase activity; and be resistant to phage infection.

The accomplishment of these requirements improves the safety of sausage and extends its shelf life, which contributes to a substantial reduction in processing time and a sausage with standardized quality. When long-ripened high-quality sausages are produced, a combination of LAB for pH drop and *Kocuria* for nitrate/nitrite reduction is used. A summary of the metabolic properties and consequences for quality are summarized in Fig. 5.4.

Production, Quality Control and Application of Starter Cultures

The production of starter cultures is based on its cost and benefits in terms of biomass, enzymatic activity profile, suitability to freeze-drying and high stability for medium- to long-term storage. Cells are usually collected toward the end of the exponential growth phase or the beginning of the stationary phase, cooled and concentrated either by centrifugation or ultrafiltration.

The cultures are supplied frozen or freeze-dried. In the case of frozen storage, cultures are kept with protective agents at –20 to –40C as a short-term (2–3 months) method of preservation. Longer storage times would need temperatures as low as -196C. The freeze-dried cultures consist in a powder with less than 3% residual moisture. The count usually shows little reduction as compared with the original culture. If correctly packaged and in the absence of air, the viability of bacteria can remain unaltered for 5–6 months at -20C. At room temperature the activity of the cultures declines more rapidly, but enough cells will survive. Mold cultures are supplied as freeze-dried spore suspensions (the spore suspend readily in water) or as lyophilized powder and yeast cultures as freeze-dried cells.

All starter cultures must be routinely surveyed for absence of pathogenic or spoilage microorganisms as well as toxic contaminant compounds. The quality control must also attend several aspects like the fermentative activity, the total count of starter organisms, the stability of genetic characteristic and the declared shelf life. Other controls include tests of acidification for lactic acid bacteria and nitrate reduction for *Micrococcaceae*. The effectiveness of the culture will depend on the growth rate and its metabolic activity in the conditions of sausage substrate.

Bacteriophages, or phages, can proliferate in a bacterial cell, interfering with its metabolism to produce phages and result in the lysis of the cell and release of generated phages. The infection process follows several steps: the adsorption of the phage onto the bacterial cell, the penetration through the cell wall, the intracellular growth and the lysis of the cell and release of the phages. Phages occur in a wide variety of species but only act on certain species or strains. For instance, sensitivity to bacteriophages and presence of prophages has been described in *L. sakei* and *L. plantarum* (Nes *et al.* 1988; Leuschner *et al.* 1993) but never in pediococci. So, it is important to select bacterial strains for resistance to phages. In case of contamination, phages can be inactivated by heat treatment, disinfection (like hypochloric acid) and UV radiation. In fermented

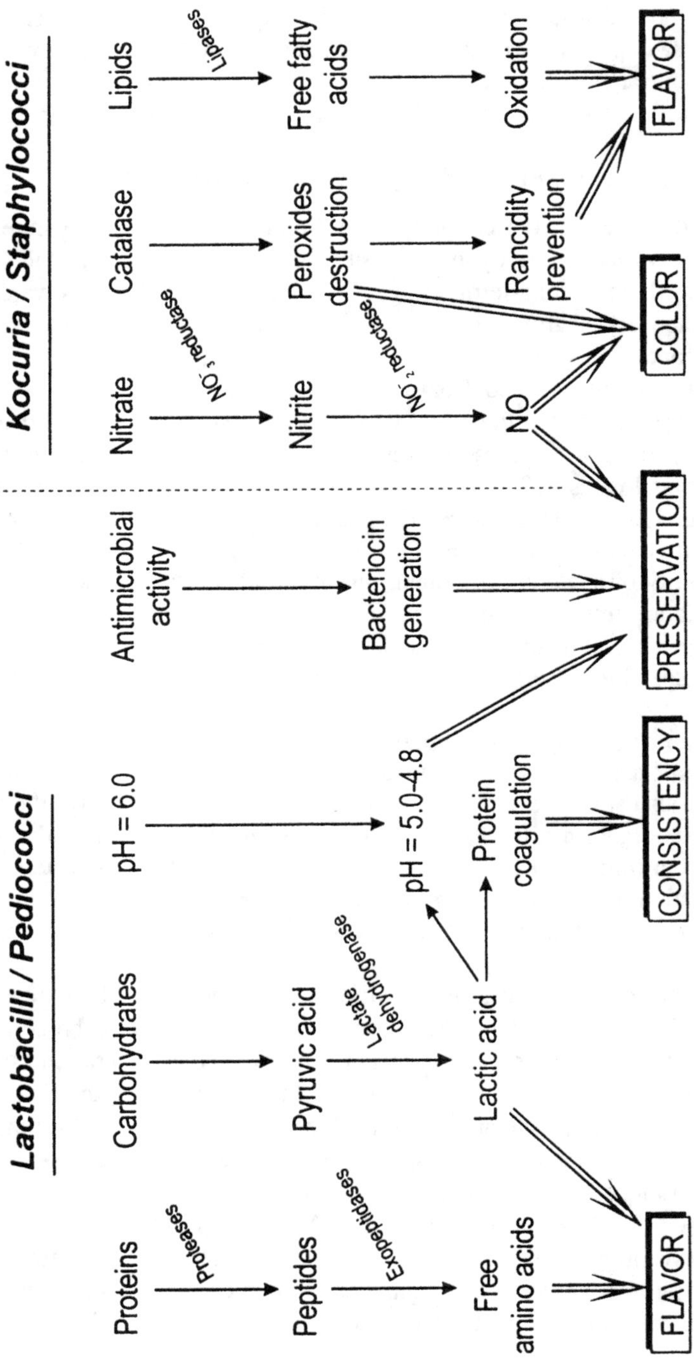

FIG. 5.4. EFFECT OF STARTER CULTURES BASED ON THEIR METABOLIC PROPERTIES

meats, the phage action is restricted because the solid matrix retards diffusion, and furthermore, starter cultures grow in niches as micropopulations isolated one from each other (Monfort, personal communication 2001).

The important feature of freeze-dried cultures is that the microorganism needs an extended period of time, which depends on the type of microorganism, choice of cryoprotective agent, conditions of resuspension, etc., for resuscitation before achieving full activity. So, they must be reconstituted in water before their addition to the meat mixture. The cultures are usually activated by keeping them at room temperature for 18–24h before their addition to the mix. The highest levels are inoculated for the production of rapid fermented sausages. Cultures of lactic acid bacteria and *Kocuria/Staphylococcus* strains are usually combined in order to have a good acidification profile with catalase and nitrate reductase activity. Molds and yeasts are only applied on the outer surface of the sausage by dipping into a suspension containing the cultures. The spores suspend readily in water. The suspension may be also sprayed on the surface.

REFERENCES

BOCKELMAN, W. 1995. The proteolytic system of starter and non-starter bacteria: Components and their importance for cheese ripening. Int. Dairy J. *5*, 977–994.

CASSENS, R.G. 1997. Composition and safety of cured meats in the USA. Food Chem. *59*, 561–566.

DEMEYER, D.I. 1992. Meat fermentation as an integrated process. *In* New Technologies for Meat and Meat Products, J.M. Smulders, F. Toldrá, J. Flores and M. Prieto (eds.), pp. 21–36. Audet, Nijmegen, The Netherlands.

DEMEYER, D. *et al.* 2000. Control of bioflavour and safety in fermented sausages: first results of a European project. Food Research Int. *33*, 171–180.

DIERICK, N., VANDEHERCKHOVE, P. and DEMEYER, D. 1974. Changes in nonprotein nitrogen compounds during dry sausage ripening. J. Food Sci. *39*, 301–304.

EVERSON, C.W., DANNER, W.E. and HAMMES, P.A. 1970. Bacterial starter cultures in sausage products. J. Agric. Food Chem. *18*, 570–571.

FADDA, S., SANZ, Y., VIGNOLO, G., ARISTOY, M-C., OLIVER, G. and TOLDRÁ, F. 1999a. Hydrolysis of pork muscle sarcoplasmic proteins by *Lactobacillus curvatus* and *Lactobacillus sake*. Appl. Environ. Microbiol. *65*, 578–584.

FADDA, S., SANZ, Y., VIGNOLO, G., ARISTOY, M-C., OLIVER, G. and TOLDRÁ, F. 1999b. Characterization of muscle sarcoplasmic and myofibrillar protein hydrolysis caused by *Lactobacillus plantarum*. Appl. Environ. Microbiol. *65*, 3540–3546.

FLORES, J. and TOLDRÁ, F. 1993. Curing: Processes and applications. *In* Encyclopedia of Food Science, Food Technology and Nutrition, R. Macrae, R. Robinson, M. Sadle, G. Fullerlove (eds.), pp. 1277–1282. Academic Press, London, U.K.

FRANSEN, N.G., O'CONNELL, M.B. and ARENDT, E.K. 1997. A modified agar medium for the screening of proteolytic activity of starter cultures for meat fermentation purposes. Int. J. Food Microbiol. *36*, 235–239.

HAMMES, W.P., BANTLEON, A. and MIN, S. 1990. Lactic acid bacteria in meat fermentation. FEMS Microbiol. Reviews *87*, 165–174.

HAMMES, W.P. and KNAUF. 1994. Starters in the processing of meat products. Meat Sci. *36*, 155–168.

HAMMES, W.P. and TICHACZEK, P.S. 1994. The potential of lactic acid bacteria for the production of safe and wholesome food. Z. Lebensm. Unters. Forsch. *198*, 193–201.

HUGAS, M. and MONFORT, J.M. 1997. Bacterial starter cultures for meat fermentation. Food Chem. *59*, 547–554.

IGENE, J.O., YAMAUCHI, K., PEARSON, A.M. and GRAY, J.L. 1985. Mechanisms by which nitrite inhibits the development of warmed-over flavor in cured meat. Food Chem. *18*, 1–18.

JOHANSSON, G., BERDAGUÉ, J-L., LARSSON, M., TRAN, N. and BORCH, E. 1993. Lipolysis, proteolysis and formation of volatile components during ripening of a fermented sausage with *Pediococcus pentosaceus* and *Staphylococcus xylosus*. Meat Sci. *38*, 203–218.

KRÖCKEL, L. 1995. Bacterial fermentation of meats. *In* Fermented Meats, G. Campbell-Platt and P.E. Cook (eds.), pp. 69. Chapman & Hall, London, U.K.

LEHNINGER, A.L., NELSON, D.L. and COX, M.M. (eds.). 1993. Principles of Biochemistry. 2nd Ed., pp. 511. Worth Publishers, New York, NY.

LEISTNER, L., GEISEN, R. and BOECKLE, B. 1991. Possibilities and limits to genetic change in starter cultures and protective cultures. Fleischwirts. *71*, 682–683.

LEISTNER, F. 1992. The essentials of producing stable and safe raw fermented sausages. *In* New Technologies for Meat and Meat Products, J.M. Smulders, F. Toldrá, J. Flores and M. Prieto (eds.), pp. 1–19. Audet, Nijmegen, The Netherlands.

LEUSCHNER, R.G.K., ARENDT, E.K. and HAMMES, W.P. 1993. Characterization of a virulent *Lactobacillus sake* phage PWH2. Appl. Microbiol. Biotechnol. *39*, 617–621.

LÜCKE, F.K. 1985. Fermented sausages. *In* Microbiology of Fermented Foods, B.J.B. Wood (ed.) Vol. 2, pp. 41–83. Elsevier Applied Science, London, U.K.

LÜCKE, F-K. and HECHELMAN, H. 1987. Starter cultures for dry sausages and raw ham. Composition and effect. Fleischwirtsch. *67*, 307–314.

LÜCKE, F.K. 1992. Prospects for the use of bacteriocins against meat-borne pathogens. *In* New Technologies for Meat and Meat Products, J.M. Smulders, F. Toldrá, J. Flores and M. Prieto (eds.), pp. 37–52. Audet, Nijmegen, The Netherlands.

MOLLY, K., DEMEYER, D., JOHANSSON, G., RAEMAEKERS, M., GHISTELINK, M. and GEENEN, I. 1997. The importance of meat enzymes in ripening and flavour generation in dry fermented sausages. First results of an European project. Food Chem. *59*, 539–545.

MONNET, V., CONDON, S., COGAN, T.M. and GRIPON, J.C. 1996. Metabolism of starter cultures. *In* Dairy Starter Cultures, T.M. Cogan and J-P. Accolas (eds.), pp. 47–91. VCH, New York, NY.

MONTEL, M.C., TALON, R., BERDAGUE, J.L. and CANTONNET, M. 1993. Effects of starter cultures on the biochemical characteristics of French dry sausages. Meat Sci. *35*, 229–240.

MONTEL, M.C., SERONINE, M.P., TALON, R. and HEBRAUD, M. 1995. Purification and characterization of a dipeptidase from *Lactobacillus sake*. Appl. Environ. Microbiol. *61*, 837–839.

NES, I.F., BRENDEHANG, J. and HUSBY, K.O. 1988. Characterization of the bacteriophage B2 of *Lactobacillus plantarum* ATCC 8014. Biochimie *70*, 423–427.

NIINIVAARA, F.P., POHJA, M.S. and KOMULAINEN, S.E. 1964. Some aspects about using bacterial pure cultures in the manufacture of fermented sausages. Food Technol. *18*, 147–153.

ORDOÑEZ, J.A., HIERRO, E.M., BRUNA, J.M. and DE LA HOZ, L. 1999. Changes in the components of dry-fermented sausages during ripening. Crit. Rev. Food Sci. Nutr. *39*, 329–367.

PRITCHARD, G.G. and COOLBEAR, T. 1993. The physiology and biochemistry of the proteolytic system in lactic acid bacteria. FEMS Microb. Reviews *12*, 179–206.

ROCA, M. and INCZE, K. 1990. Fermented sausages. Food Reviews Int. *6*, 91–118.

RONCALÉS, P., AGUILERA, M., BELTRÁN, J.A., JAIME, I. and PEIRO, J.M. 1991. The effect of natural and artificial casing on the ripening and sensory quality of a mould-covered dry sausage. Int. J. Food Sci. Technol. *26*, 83–89.

SAMELIS, J., STAVROPOULOS, S., KAKOURI, A. and METAXOPOULOS, J. 1994. Quantification and characterization of microbial populations associated with naturally fermented Greek dry salami. Food Microbiol. *11*, 447–460.

SANZ, B., SELGAS, D., PAREJO, I. and ORDÓÑEZ, J.A. 1988. Characteristics of lactobacilli isolated from dry fermented sausages. Int. J. Food Microbiol. *6*, 199–205.

SANZ, Y. and TOLDRÁ, F. 1997. Purification and characterization of an aminopeptidase from *Lactobacillus sakei*. J. Agric. Food Chem. *45*, 1552–1558.

SANZ, Y., MULHOLLAND, F. and TOLDRÁ, F. 1998. Purification and characterization of a tripeptidase from *Lactobacillus sake*. J. Agric. Food Chem. *46*, 349–353.

SANZ, Y., FADDA, S., VIGNOLO, G., ARISTOY, M-C., OLIVER, G. and TOLDRÁ, F. 1999a. Hydrolytic action of *Lactobacillus casei* CRL 705 on pork muscle sarcoplasmic and myofibrillar proteins. J. Agric. Food Chem. *47*, 3441–3448.

SANZ, Y., FADDA, S., VIGNOLO, G., ARISTOY, M-C., OLIVER, G. and TOLDRÁ, F. 1999b. Hydrolysis of muscle myofibrillar proteins by *Lactobacillus curvatus* and *Lactobacillus sake*. Int. J. Food Microbiol. *53*, 115–125.

SANZ, Y. and TOLDRÁ, F. 2002. Purification and characterization of an X-prolyl-dipeptidyl peptidase from *Lactobacillus sakei*. Appl. Environ. Microbiol. (In press).

STAHNKE, L.H. 1995. Aroma components from dried sausages fermented with *Staphylococcus xylosus*. Meat Sci. *38*, 39–53.

STRAUB, B.W., KICHERER, M., SCHILCHE, S.M. and HAMMES, W.P. 1995. The formation of biogenic amines by fermentation organisms. Z. Lebensm. Unters. Forsch. *201*, 79–82.

TALON, R., WALTER, D., CHARTIER, S., BARRIERE, C. and MONTEL, C. 1999. Effect of nitrate and incubation conditions on the production of catalase and nitrate reductase by staphylococci. Int. J. Food Microbiol. *52*, 47–56.

TOLDRÁ, F., SANZ, Y. and FLORES, M. 2001. Meat Fermentation Technology. *In* Meat Science and Applications, Y.H. Hui, H. Shorthose, O. Young, M. Koohmaraie and R.W. Rogers (eds.), pp. 537–561. Marcel Dekker, New York.

TOLDRÁ, F. and VERPLAETSE, A. 1995. Endogenous enzyme activity and quality for raw product processing. *In* Composition of Meat in Relation to Processing, Nutritional and Sensory Quality, K. Lündstrom, I. Hansson and E. Winklund (eds.), pp. 41–55. ECCEAMST, Uppsala, Sweden.

WONG, D.W.S. 1995. Lipolytic enzymes. *In* Food Enzymes. Structure and Mechanism, pp. 170–211. Chapman and Hall, New York.

CHAPTER 6

CHARACTERIZATION OF PROTEOLYSIS

PROTEOLYSIS IN DRY-CURED HAM

Proteolysis constitutes one of the most important group of biochemical reactions in the generation of flavor and/or flavor precursors during the processing of dry-cured ham. Proteolysis in dry-cured ham is mainly attributed to the endogenous enzymatic systems in view of the low microbial counts found inside the hams, the low microbial enzyme activity levels and the difficult conditions for microbial growth (Molina and Toldrá 1992). The pH, salt concentration and low moisture content appears to be limiting factors for bacterial growth.

Proteolysis has a high impact in the quality of dry-cured ham for several reasons: direct contribution to texture by breakdown of the myofibrillar proteins responsible for muscle network, generation of peptides and free amino acids with direct influence on taste and generation of flavor precursors, such as free amino acids, that will act as substrates for further reactions contributing to flavor. In general, the most important proteolytic changes have been observed in hams with longer processing time and less salt content (Toldrá and Flores 1998). Most of the recent research and reports have been focused on the muscle enzyme systems as an effective way to understand the process and control it to optimize the final quality.

Action of Muscle Proteases

The progress of proteolysis in dry-cured ham can vary depending on the type of product, the amount of endogenous proteolytic enzymes and the specific process conditions. A description of main muscle proteolytic enzymes is given in Chap. 2. In general, proteolysis has the following stages (Fig. 6.1): an initial breakdown of major myofibrillar proteins by calpains and cathepsins and the formation of protein fragments and intermediate-size polypeptides resulting from the hydrolysis, subsequent degradation of these polypeptides to small peptides by di- and tripeptidylpeptidases and the final generation of free amino acids resulting from the action of dipeptidases, aminopeptidases and carboxypeptidases. As mentioned above, the action of microbial proteases has also been studied. No proteolytic activity was detected when microbial proteases from *P. pentosaceus* and *S. xylosus*, the most important microorganisms found in dry-cured ham, were incubated with myofibrillar and sarcoplasmic proteins obtained in sterile conditions. Only some aminopeptidase activity was detected although in almost negligible amounts due to the low numbers of microorganisms usually found inside the hams.

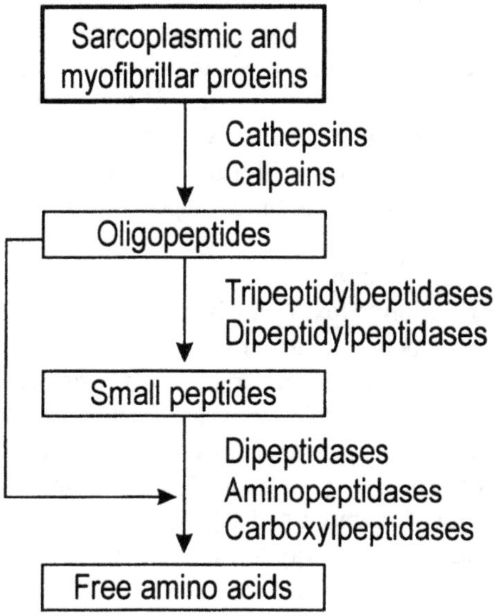

FIG. 6.1. GENERAL SCHEME SHOWING MAJOR STEPS IN PROTEOLYSIS DURING THE PROCESSING OF DRY-CURED HAM

Muscle Proteinases. The main muscle proteinases (endopeptidases) are cathepsins and calpains. Cathepsins B, D, H and L are active at acid pH values, located in the lysosome and small in size, in the range of 20–40 Kda. Cathepsins B, H and L, which are cysteine proteinases, are active through the entire dry-curing process (Fig. 6.2 (A)) and show good stability since a residual 5–10% activity is usually found even after 15 months of process (Toldrá and Etherington 1988; Toldrá *et al.* 1993). Cathepsin D, an aspartic proteinase, remains active up to six months of processing. *In vitro* assays have shown the ability of cathepsins to degrade different myofibrillar proteins, such as titin, myosin heavy chains, actin, tropomyosin and troponins T and I by cathepsins D and L and myosin and actin by cathepsin B (Goll *et al.* 1983). On the other hand, cathepsin H has particular properties showing both endo and aminopeptidase activity.

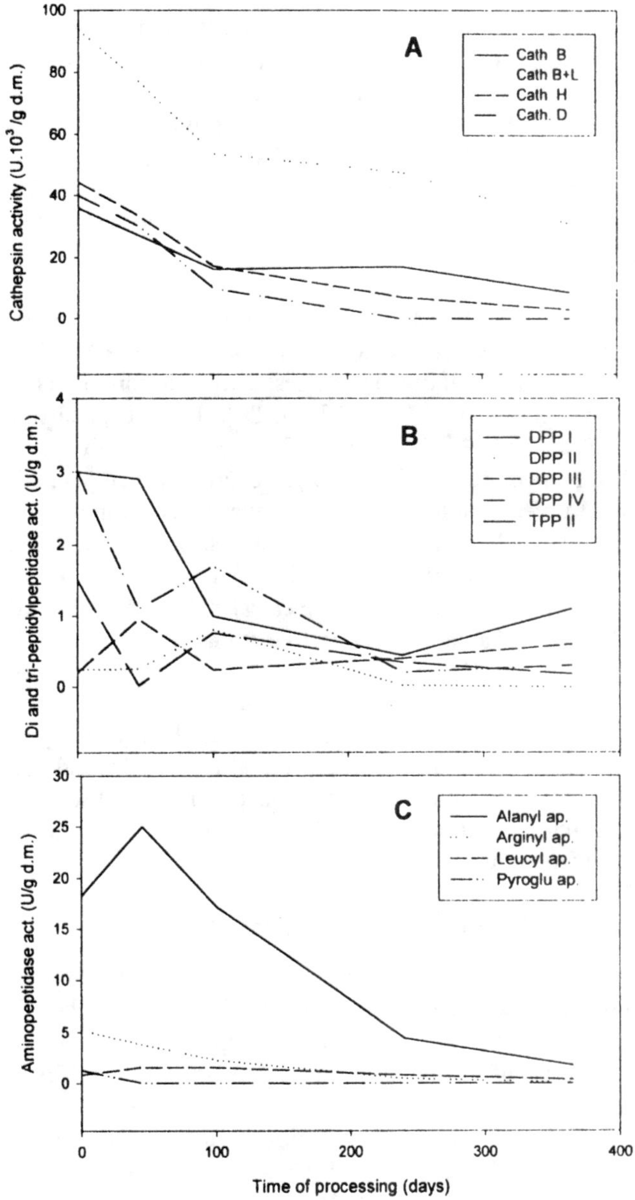

FIG. 6.2. EVOLUTION OF MUSCLE ENZYMES ACTIVITY IN THE MUSCLE *BICEPS FEMORIS* DURING THE PROCESSING OF SERRANO TYPE DRY-CURED HAM

(A) Cathepsin B, B+L and H; (B) Dipeptidylpeptidase I, II, III, IV and Tripeptidylpeptidase II; (C) Alanyl, Arginyl, Leucyl and Pyroglutamyl aminopeptidases.

(Adapted from Toldrá *et al.* 2000 and Sentandreu and Toldrá 2001)

Calpains I and II, cystein endopeptidases, are located in the cytosol and in the Z-disc region and have optimal activity at neutral pH, around 7.5. They require 50–70 μM of Ca^{2+} and 1–5 mM of Ca^{2+}, respectively, for activation. Calpains are able to degrade titin, nebulin, C-protein, troponins T and I, tropomyosin, filamin, desmin and vinculin (Koohmaraie 1994) but are unable to degrade myosin, actin, α-actinin and troponin C. Calpains stability is rather poor, since its activity is lost in 10–14 days, just after the salting stage (Rosell and Toldrá 1996).

Muscle Exopeptidases. These enzymes are involved in latter stages of proteolytical degradation. Tri- and dipeptidylpeptidases (TPP and DPP, respectively) generate tri- and dipeptides, respectively, from the N-terminal of proteins and polypeptides. TPP I, DPP I and DPP II are located in lysosomes and have optimal acid pH. TPP II and DPP III are located in the cytosol and DPP IV is linked to membranes. These enzymes have optimal pH in the neutral-basic range and are very stable, being active even after 15 months of process, 8 months for DPP II (Sentandreu and Toldrá 2001). Arginyl (RAP), leucyl (LAP), alanyl (AAP), methionyl (MAP) and pyroglutamyl (PGAP) aminopeptidases are located in the cytosol, active at neutral pH (RAP and AAP) or basic (LAP and PGAP) pH and able to generate free amino acids from the N-terminal of proteins and peptides. Alanyl and methionyl aminopeptidases have a broad substrate specificity while arginyl aminopeptidase (or aminopeptidase B) is chloride activated and hydrolyzes basic amino acids. Leucyl and pyroglutamyl aminopeptidases are present in porcine muscle at low levels, and their optimal pH is far from that in ham. Aminopeptidases have shown good activity along the processing of dry-cured ham and a high stability with significant activity still recovered after 15 months of process.

Protein Breakdown, Peptide Generation/Breakdown

The action of calpains and cathepsins in a short process like meat aging is associated with the fragmentation of myofibrils through the Z-disc, the hydrolysis of desmin, titin and nebulin and the appearance of two polypeptides with molecular masses of 95 and 30 KDa (Koohmaraie 1994). The result is an intense breakdown of proteins and an increase in tenderness. The action of muscle proteinases is even more intense during the long processing of dry-cured ham, up to 15 months (Toldrá *et al.* 1993). The electrophoretic patterns of muscle sarcoplasmic proteins (Fig. 6.3) and myofibrillar proteins (Fig. 6.4) reveal important changes for most proteins along the processing of dry-cured ham. Myosin heavy chains and light chains 1 and 2, troponins C and I disappear while several fragments with 150, 95 and 16 Kda and in the ranges 50–100 KDa and 20–45 KDa are formed. In general, hydrolysis and unsolubility are more

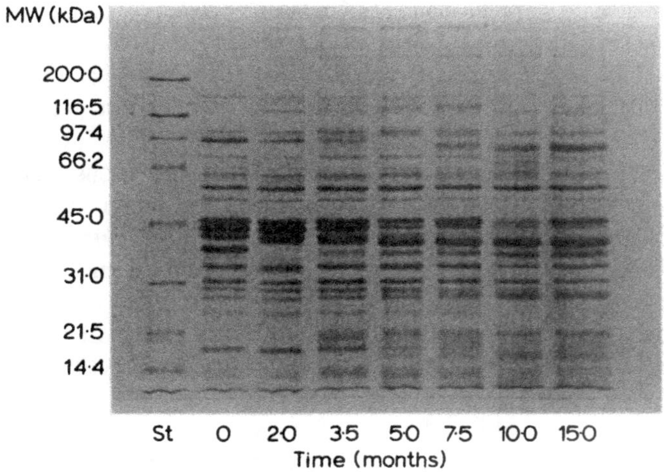

FIG. 6.3. SDS-PAGE PATTERNS OF WATER SOLUBLE PROTEINS AT DIFFERENT STAGES OF THE PROCESSING OF SERRANO DRY-CURED HAM
(From J. Sci. Food Agric. 62, Toldrá, Rico and Flores. Cathepsin B, D, H and L activities in the processing of dry-cured ham, 157–161, 1993, Copyright Society of Chemical Industry. Reproduced with permission granted by John Wiley & Sons Ltd. on behalf of the SCI)

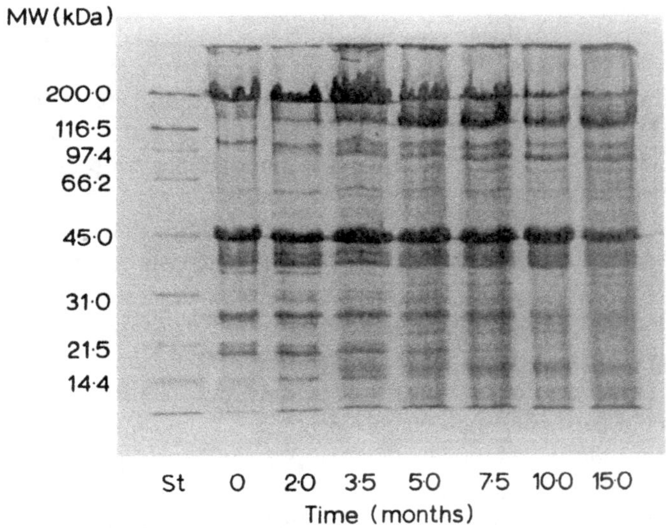

FIG. 6.4. SDS-PAGE PATTERNS OF MYOFIBRILLAR PROTEINS AT DIFFERENT STAGES OF THE PROCESSING OF SERRANO DRY-CURED HAM
(From J. Sci. Food Agric. 62, Toldrá, Rico and Flores. Cathepsin B, D, H and L activities in the processing of dry-cured ham, 157–161, 1993, Copyright Society of Chemical Industry. Reproduced with permission granted by John Wiley & Sons Ltd. on behalf of the SCI)

intense in myofibrillar proteins than in sarcoplasmic proteins. These changes are similar in Spanish Serrano and French Bayonne hams (Toldrá *et al.* 1992; Monin *et al.* 1997). Marked ultrastructural changes have also been observed by both scanning and electron microscopy. Some of the most important changes are in the breakdown, and sometimes weakening, of the Z-line, as well as in important damages through the fibers (Fig. 6.5). Fiber disruptions appear more frequently at the end of salting, these being the changes similar in the muscles *Semimembranosus* and *Biceps femoris* (Monin *et al.* 1997). In view of the proteinase properties and stabilities, the action of calpains would be restricted to the initial days, cathepsin D to the first 6 months and cathepsins B, L and H would act during the whole dry-curing process. In any case, cathepsins B and L would play the major role due to their optimal pH, good activity and high stability.

Some problems related with texture and sensory perception are associated with an excess of proteolysis. The excess of proteolysis is due to the breed types and/or ages that have a marked influence on some enzymes or just a higher level of cathepsin activity (Chap. 10). The result is poor firmness, which can be associated with poor ratings by sensory panelists and consumers. This softness has been correlated with a high residual cathepsin B activity and low salt content (Parolari *et al.* 1994) or to a high residual cathepsin B+L activity (García-Garrido *et al.* 2000). The excessive accumulation of peptides and free amino acids may give unpleasant tastes.

The evolution of peptides, resulting from protein breakdown during the processing of dry-cured ham, has been analyzed by capillary electrophoresis gel filtration chromatography and reverse-phase HPLC (Rodríguez-Nuñez *et al.* 1995). Numerous peptides are detected, being those of higher size (2,700–4,500 Da) generated during post-salting and early ripening and then hydrolyzed to new peptides of smaller size (below 2,700 Da). An example of a peptide mapping is shown in Fig. 6.6. An enhancement in the concentration of peptides, when low salt levels are added, confirms the findings showing cathepsin inhibition by salt (Rico *et al.* 1991). So, the accumulation of peptides is depressed by salt level (Martín *et al.* 1998). Smaller peptides and free amino acids accumulate in the ham as final products of the proteolytic chain. The most important peptidase appears to be DPP I due to its optimal pH, activity and stability. Contribution of DPP IV would be lower, and the other peptidases would play a minor role. Recently, a tripeptide containing Glu, Val and Asp and some dipeptides like Ile-Val, Leu-Glu, Ile-Asp, Ala-Met, Gly-Glu, Glu-Arg, Pro-Leu, Gly-Ser, Asp-Val and Ser-Lys have been isolated and identified at the end of the process (Sentandreu *et al.* 2002).

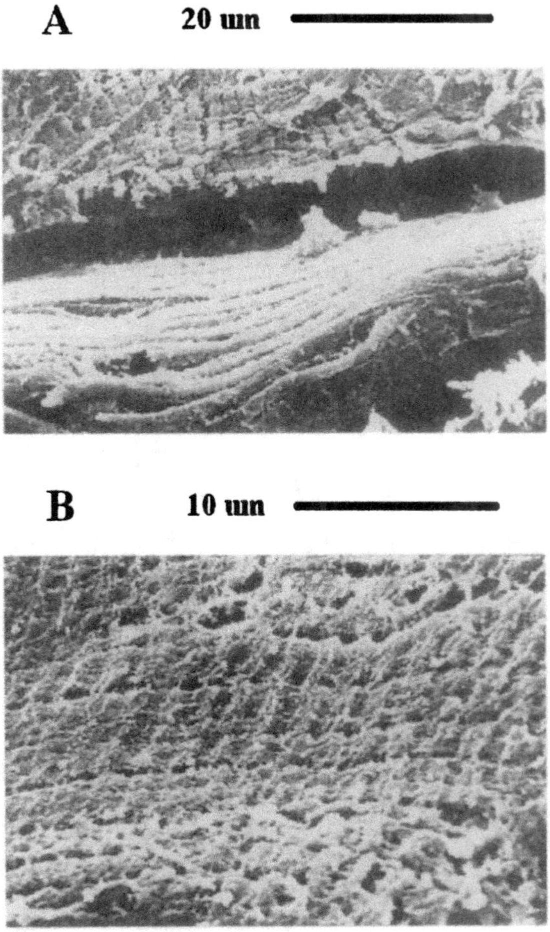

FIG. 6.5. SCANNING ELECTRON MICROGRAPH OF MUSCLE FIBERS IN 8 MONTHS DRY-CURED HAM
(A) General view of the fibers, (B) Specific details of the structure.
(Toldrá and Voyle 1988, Unpublished)

Generation of Free Amino Acids

As a consequence of the intense proteolysis experienced during dry curing, there is an incredible high generation of free amino acids along the processing of dry-cured ham. Alanine, leucine, valine, arginine, lysine, glutamic and aspartic acids are some of the generated amino acids in higher amounts. The final concentrations depend on the length of the process and type of ham (Fig. 6.7). For instance, the content in free amino acids in Parma hams increases

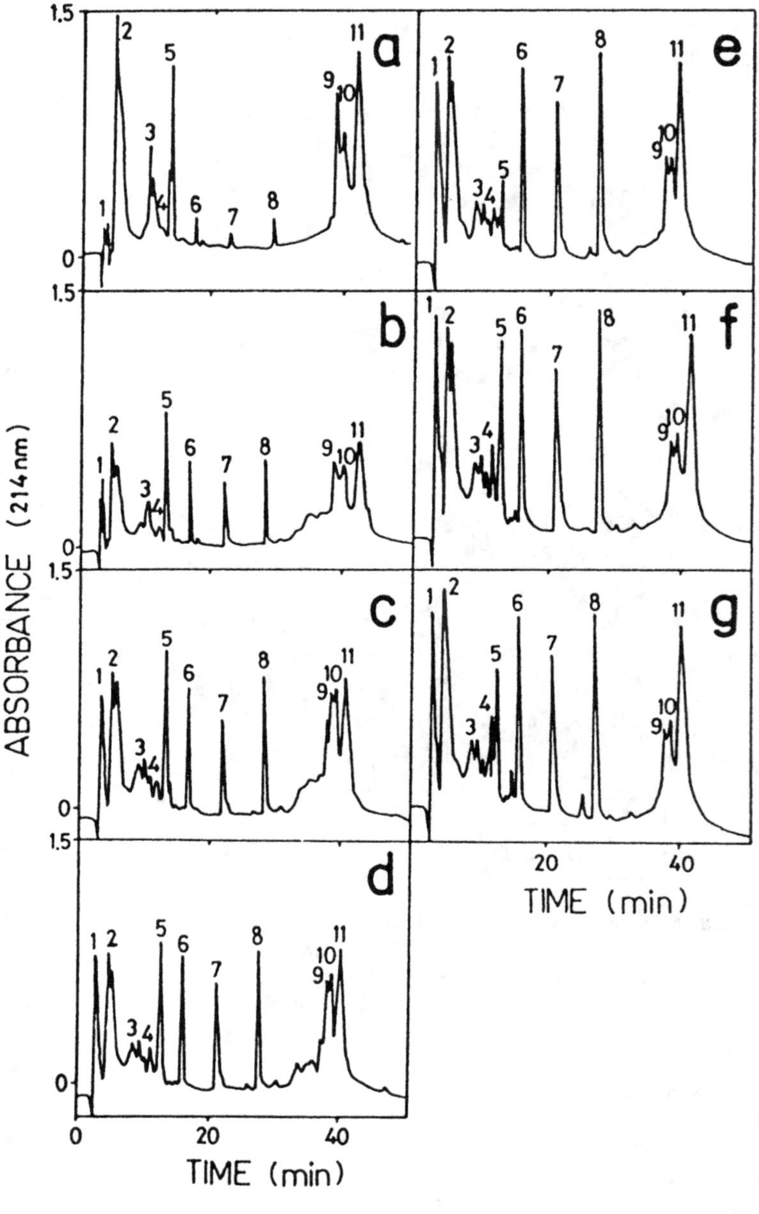

FIG. 6.6. PEPTIDE MAPPING ALONG THE PROCESSING OF SERRANO TYPE DRY-CURED HAM

(From Food Chem. 53, Rodríguez-Nuñez, Aristoy and Toldrá, Peptide Generation in the Processing of Dry-Cured Ham, 187–190, 1995, with permission from Elsevier Science)

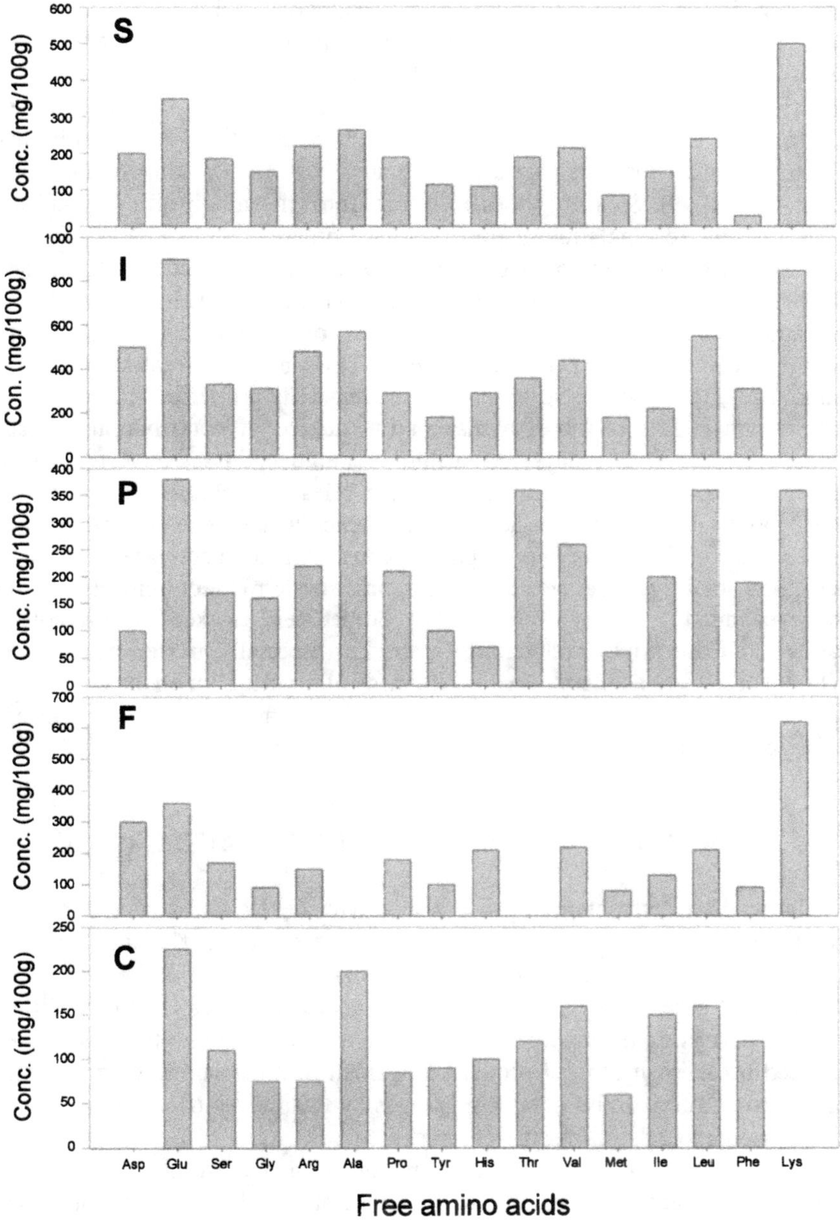

FIG. 6.7. TYPICAL PROFILE OF FREE AMINO ACIDS GENERATED IN SERRANO (S), IBERIAN (I), PARMA (P), FRENCH (F) AND COUNTRY STYLE (C) DRY-CURED HAMS (Adapted from Toldrá and Flores 1998)

particularly over the 9–12 month period (Schivazappa *et al.* 1995). On the other hand, Spanish hams show the highest rate of generation taking place at up to 240 days, which coincides with intense protein breakdown and maximal enzyme activity (Fig. 6.8). Afterward, the generation trend is slower due to the reduction in enzyme activity and further reactions to form other compounds (Toldrá *et al.* 2000). This decrease has been observed after 179 days in French dry-cured hams (Buscailhon *et al.* 1994) and after 420 days in Iberian hams (Ruiz *et al.* 1999). Small peptides resulting from tri- and dipeptidylpeptidases action are still accumulating at the end of the process.

The highest concentrations of free amino acids are detected in Iberian ham, which has a long processing time (more than 24 months). The lowest amount of free amino acids are in country ham because of its shorter processing time (Toldrá and Flores 1998). Based on the increase in the concentration of free amino acids, the substrate specificity and the specific properties for activity of each enzyme, it is possible to discern the degree of contribution of each aminopeptidase. Thus, alanyl and methionyl aminopeptidases appear to be the most important enzymes contributing to the release of almost all free amino acids, followed by arginyl aminopeptidase, which would generate arginine and lysine. The contribution of leucyl and pyroglutamyl aminopeptidases, even with their good stability, is restricted due to optimal basic pH requirements and the low level found in muscle (Toldrá 1998). Sometimes, an excess of proteolysis results in an excessive production of peptides (especially in the range 26–87 KDa) that are responsible for the development of a white film on the cut surface of vacuum-packaged slices (Toldrá *et al.* 1990), or free amino acids that lead to the formation of visible white crystals of tyrosine.

PROTEOLYSIS IN DRY-FERMENTED SAUSAGES

In modern fermentation processes, flavor development and consistent product quality have become as important as preservation. The application of our increasing knowledge of the proteolytic enzymes of lactic acid bacteria may be useful for accelerated ripening of dry-fermented sausages, flavor modifications and correcting the nature of some defects in taste such as bitterness. A well balanced breakdown of meat proteins into small peptides and free amino acids is necessary. They can be flavor compounds by themselves or act as precursors of flavor compounds during the process.

As in the case of dry-cured ham, proteolysis also constitutes an important group of biochemical reactions in the generation of flavor and/or flavor precursors during the processing of dry-fermented sausages (Toldrá 1992). The degree of contribution of either endogenous proteases or those of microbial origin either naturally present in the product or added as starter cultures will mainly depend on the raw materials, type of product and processing conditions (Demeyer 1992; Molly *et al.* 1997).

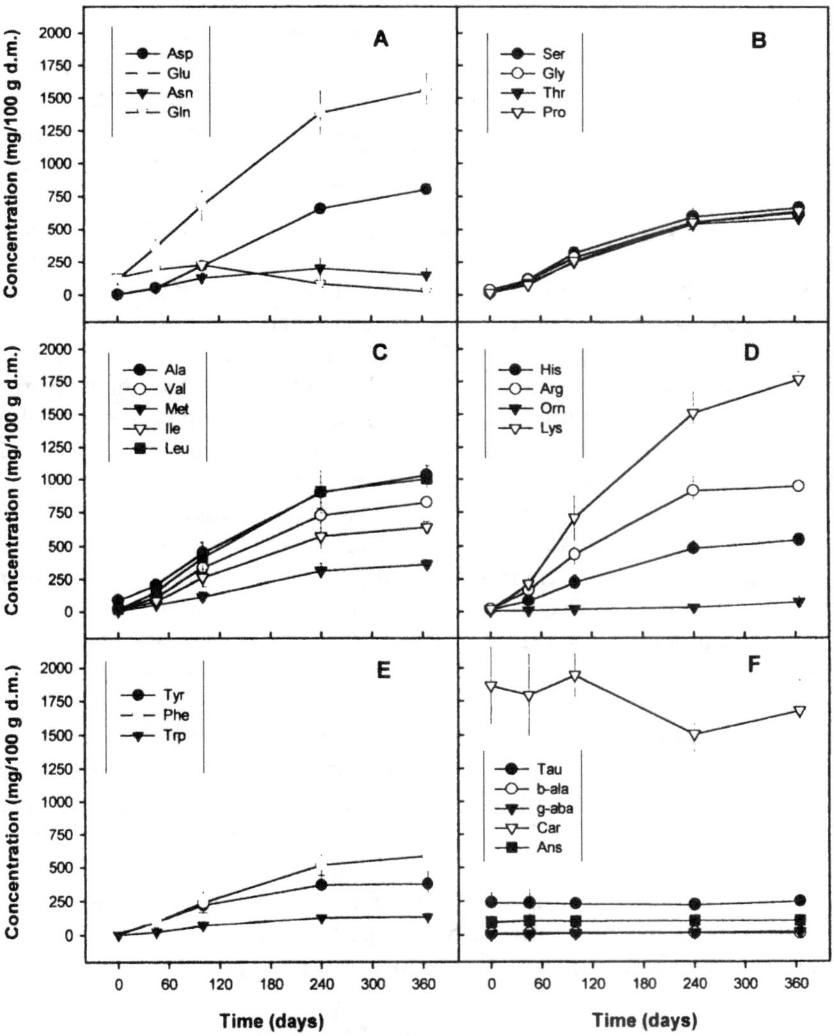

FIG. 6.8. EVOLUTION OF FREE AMINO ACIDS AND NATURAL DIPEPTIDES ALONG THE PROCESSING OF SERRANO TYPE DRY-CURED HAM
(From Food Res. Int. 33, Toldrá, Aristoy and Flores, Contribution of Muscle Aminopeptidases to Flavor Development in Dry-Cured Ham, pp. 181–185, 2000, with permission from Elsevier Science)

Action of Muscle and Microbial Proteases

The proteolytic enzymes are important for both the growth of bacteria in meat and for the contribution they make to the flavor and texture of the product. In fact, the importance of the proteolytic system has stimulated research in this area during the last decade, especially in Lactococci, due to its importance for the dairy industry. Several proteases have been purified and characterized, and the properties of the transport system of amino acids and peptides have been studied in Lactococci. The biochemical and genetic characteristics of the components of the proteolytic system and the rapid development of techniques for the genetic modification of strains will allow the engineering of strains for specific purposes by adding or removing specific proteolytic properties.

Meat is a complex system to study the proteolytic process because it is a relatively high set of proteins. Sausages produced with starter addition show excellent commercial appearance, dry more homogeneously, without rind, and develop a good color and texture. However, consumers tend to prefer sausages manufactured without starter and dried at mild temperatures due to their flavor. Manufacturing conditions, especially the pH drop profile, which may change depending on the added starter (Fig. 6.9 (A)), are very important for proteolysis. Sausages with low pH are characterized by high peptide and amino acid content and a low ammonia content. However, ammonia is also produced by deaminase activity of the internal flora, and its production improves taste development since it neutralizes acidity and increases pH (Demeyer 1992).

Molds grow in aerobic environments and are found as a layer on the outer surface of the sausages. Their role seems to be related to the inhibition of fat oxidation, the production of metabolic products that contribute to taste and flavor and keep the quality of the sausages, preventing the growth of undesirable microorganisms. Fungal species isolated from dry-fermented sausages are usually nonpathogenic, although a few might be able to produce mycotoxins under appropriate conditions. The growth of controlled strains can be used as protective cultures and to avoid the development of toxinogenic molds. The inoculation with selected mold strains may aport proteases and produce a more intense proteolysis, depending on the mold strain and type of casing (Toledo *et al.* 1997). For instance, Penicillium grows intensive and in a dense form, showing an increase in nonprotein nitrogen, and Mucor strains grow superficially and stimulate the production of low molecular mass peptides.

Protein Breakdown, Peptide Generation/Breakdown

The proteins are very important for sensory attributes of the sausages like texture, flavor and color. Protein solubility is affected by fermentation for several reasons: the presence of salt (2–3% weight), the progressive pH decrease experienced during the fermentation stage as a consequence of lactic acid production by bacterial metabolism of carbohydrates and the heating/drying during the fermentation/ripening stages. The solubility of myofibrillar proteins is reduced in 50–60% and of sarcoplasmic proteins in 20–47% when heated at 37C for 12–40 h in the presence of salt and conditions of declining pH (Klement *et al.* 1973, 1974). Even though sarcoplasmic proteins lack gelling ability and are thus considered as poor contributors to structure, their salt-induced insolubilization may affect the precipitation of myofibrillar proteins, promoting the development of a better structure than with myofibrillar proteins alone

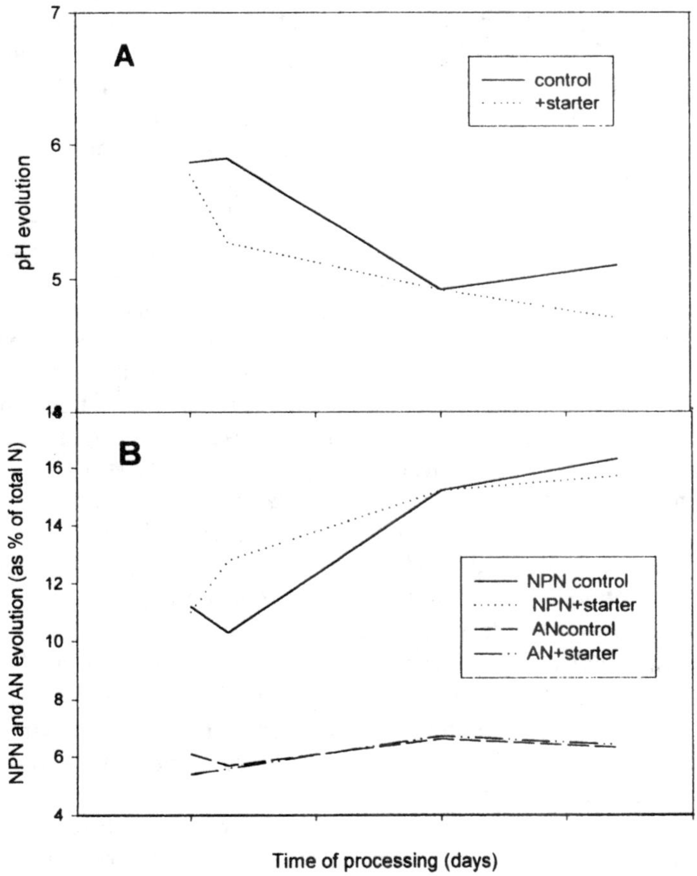

FIG. 6.9. EXAMPLE OF EVOLUTION OF pH (A), NONPROTEIN AND AMINO NITROGEN (B) DURING THE PROCESSING OF DRY-FERMENTED SAUSAGE WITHOUT ADDED STARTERS AND WITH THE ADDITION OF A STARTER CONSISTING IN A MIXTURE OF *L. SAKEI* AND *S. CARNOSUS*
(Adapted from Sanz *et al.* 1997)

(Klement *et al.* 1974). In general, the loss of solubility of sarcoplasmic proteins due to heat, pH and salt effects is below the loss in solubility of myofibrillar proteins. When there is a direct acidification by lactic acid, myofibrillar proteins appear to be more susceptible to denaturation by heat than by reduction in pH (Acton 1977).

The pH reached at the fermentation stage is very important for the action of proteinases and peptidases. When the pH drops below 5.0, the proteolytic activity by endogenous cathepsins, especially cathepsin D, is very intense (Toldrá and Verplaetse 1995). Several myofibrillar proteins are intensely degraded. This is the case of myosin and actin which are clearly degraded to fragments of 135 and 38 Kda, respectively. Other fragments of 29 and 13 Kda are also formed. When the pH drop is slight (pH above 5.0), the degradation of major proteins is not so intense. The proteolytic index, calculated as the ratio of nonprotein nitrogen to total nitrogen, varies depending on the type of product and technology. For instance, it is very high in Chorizo (Astiasarán *et al.* 1990). The relative role of endogenous muscle and bacterial enzymes in protein hydrolysis and metabolism has been investigated (Demeyer 1992; Molly *et al.* 1997). They used antibiotics and specific protease inhibitors in order to inhibit bacteria proteinases or endogenous muscle proteinases in model systems reflecting fermentation and ripening conditions. The obtained results confirmed that proteolysis was mainly due to cathepsin D (-like) enzymes, activated by pH drop, with a minor role due to bacterial proteinases. Other muscle cathepsins (B, H and L) would have a minor role in degrading actin and other fragments. The results also proved that other proteinases (trypsin-like, serin- and metallo-) were not important for proteolysis during the processing (Molly *et al.* 1997).

The proteinase activity of strains of *L. sake*, *L. plantarum*, *L. casei* and *L. curvatus*, isolated from dry-fermented sausages and assayed against myofibrillar and sarcoplasmic proteins in model fermentation systems, shows a different behavior. The proteolysis is more intense on sarcoplasmic proteins, with partial (and in some cases total) degradation of numerous proteins and the generation of an intense number of peptides (Fadda *et al.* 1999a, b; Sanz *et al.* 1999a). This is also the case for proteinases from yeasts like *Debaryomyces hansenii*, which also shows an intense proteolysis of sarcoplasmic proteins (Santos *et al.* 2001). However, myofibrillar proteins are more resistant to proteolysis with a rather poor generation of peptides (Sanz *et al.* 1999b). This would confirm the role of cathepsin D as the main proteinase acting on myofibrillar proteins.

The major increases in nonprotein nitrogen, especially peptides and small protein fragments, are produced during fermentation, heating (smoking) and early ripening (DeMasi *et al.* 1990). These increases are enhanced in the presence of starters, confirming the importance of microbial peptidases (Fig. 6.9.B). The resulting peptides have a strong influence on the final taste of the sausage, and a correct balance is of extreme importance. The generation of hydrophilic peptides during ripening is correlated with sausage taste, but, on the other hand, the generation of high amounts of hydrophobic peptides is responsible for bitter taste and off-flavors.

Some commercial proteinases, from different sources, have been assayed for enhancing the proteolysis and accelerating the process. However, most cause an increased proteolysis and an excessive softening with no sensible effect on taste. The adequate enzyme and dose must be carefully chosen, and the process conditions must be controlled. In this way, the increased protein degradation can be counteracted by the higher weight loss and produce a firmer sausage.

Generation of Free Amino Acids

The latest step in the proteolytic chain is the generation of single amino acids from peptides and protein fragments. During the ripening period, a progressive decrease in the concentration of low molecular weight peptides is observed, whereas the concentration of free amino acids increases. This hydrolysis is the result of the combined action of muscle aminopeptidases (described in Chap. 2) and microbial aminopeptidases (described in Chap. 5). The larger increases in free amino acids observed in the processing of dry-fermented sausages (Fig. 6.10) are observed for glutamic acid, alanine, arginine, valine, leucine, phenylalanine and lysine (Flores *et al.* 1998). The balance of free amino acids depends on the pH reached in the product, concentration of curing agents (salt, seasonings, nitrate/nitrite and ascorbic acid) and processing conditions (time, temperatures, dehydration) since they mainly affect the contribution of the different aminopeptidases (Sanz and Toldrá 1999). The pH reached at the fermentation stage ($pH < 5.0$) constitutes a decisive factor for muscle aminopeptidases, reducing its activity to negligible values. It also reduces the activity of microbial enzymes, although some of them are still active at those low pH values (Sanz and Toldrá 1997). Salt also exerts some inhibitory effect on most microbial exopeptidases although some of them, like AP 4, may be activated.

The generation of D-enantiomers of some amino acids, like D-alanine and D-glutamic acid, has been observed along the ripening. These compounds could be considered in the future as markers of the fermentation and used to asses the ripening state of the salamis (Baldini *et al.* 2000).

The generation of free amino acids also depends on the starter inoculation (particular strain and species). For instance, strains of *L. sakei*, *L. plantarum*, *L. casei* and *L. curvatus* show aminopeptidase activity with a wide specificity against amino acids at the N-terminal position, especially for alanine, valine and leucine. When incubated with myofibrillar and sarcoplasmic proteins, the behavior differs. So, the highest amounts of free amino acids are released from sarcoplasmic proteins. The generation of free amino acids is particularly high when incubated with *L. sakei* and *L. plantarum* and lower when incubated with *L. casei* and *L. curvatus*. However, the generation of free amino acids is rather poor when these strains are incubated with myofibrillar proteins (Fadda *et al.* 1999a, b; Sanz *et al.* 1999b).

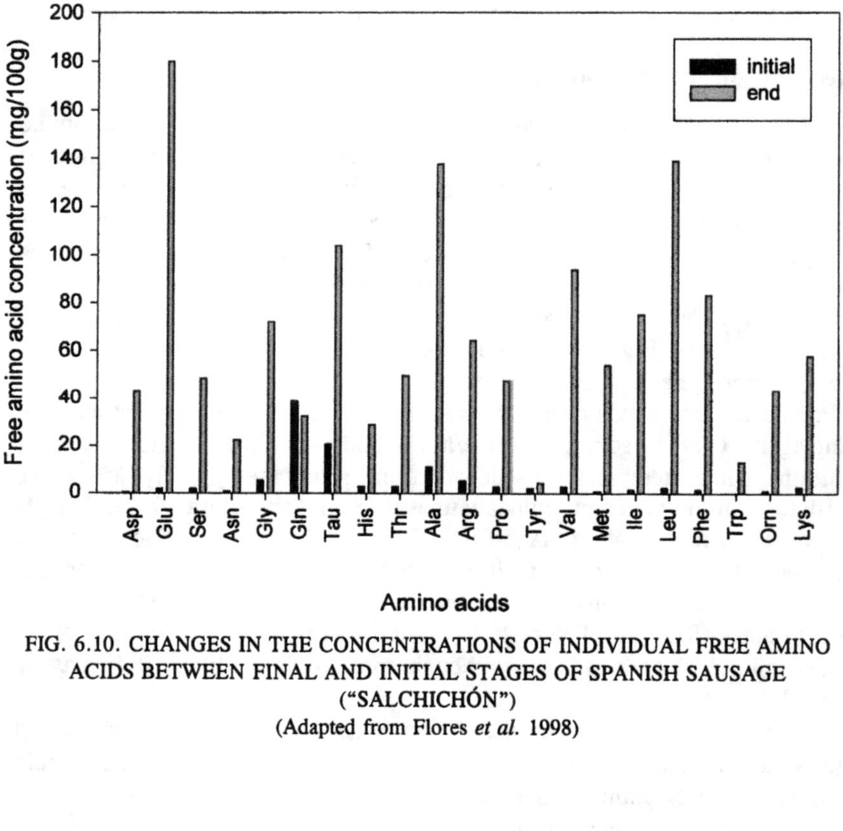

FIG. 6.10. CHANGES IN THE CONCENTRATIONS OF INDIVIDUAL FREE AMINO ACIDS BETWEEN FINAL AND INITIAL STAGES OF SPANISH SAUSAGE ("SALCHICHÓN")
(Adapted from Flores *et al.* 1998)

The influence of molds depends on the mold strain that grows on the surface and the type of casing used (Toledo *et al.* 1997). The total volatile basic nitrogen increases towards the end of the process, especially during drying, as a result of ammonia production by deamidase and deaminase activity, enzymes typically found in molds and yeasts. The neutralization of acidity produces an increase in pH and favors the action of both microbial and muscle exopeptidases.

METABOLISM OF AMINO ACIDS: AMINE GENERATION

Microorganisms exert several reactions involved with the metabolism of amino acids, such as the generation of amines. The production will depend on the availability of free amino acids, presence of microorganisms with decarboxylase activity and adequate conditions for microbial growth (Tarján and Jánossy 1978). The a_w levels and low pH, typical of dry-fermented sausages, also potentiate the amine formation. The decarboxylation of free amino acids, like histidine, tyrosine and tryptophan to respective amines histamine, tyramine and tryptamine, constitute an important risk for consumers. The relevant amines, with their corresponding precursor amino acid and ranges of content in dry-cured meat products, are given in Table 6.1. Amine content in dry-cured hams is very low, generally lower than 1 mg/100 g, and is similar to those levels found in cooked products. In the case of dry-fermented sausages, tyramine is the monoamine found at higher levels, up to 150 mg/100 g, and has been recently reported as the main amine found in sausages within 38–62% of the total amount of amines (Demeyer *et al.* 2000). In particular cases, it may reach limits as high as 10 to 80 mg/100 g of food, which is considered unsafe (Hernández-Jover *et al.* 1997). 2-phenylethylamine and tryptamine, derived from phenylalanine and tryptophan, respectively, are found at lower levels, below 3.5 mg/100 g and 8.7 mg/100 g, respectively (Maijala *et al.* 1995), even though the minimal effective oral dose for phenylethylamine is as low as 5 mg (Lücke 1985). Putrescine and cadaverine have been reported at levels below 30 mg/100 g dry matter (Vandekerckhove 1977). The amounts of histamine are very low and are far from the dose of 100 mg necessary for symptoms in humans (Ordoñez *et al.* 1999). Spermine and spermidine may reach 4.7 mg/100 g and 1.8 mg/100 g, respectively, in sausages (Maijala *et al.* 1995) although higher levels, up to 30 mg/100 g, have been reported for different types of sausages (Demeyer *et al.* 2000).

The presence of biogenic amines in dry-fermented sausages, as in other foods such as cheese and wine, may constitute a potential public hazard. Except when large quantities are consumed, amines are usually decomposed by deamination once in humans by monoamine oxidase (MAO), an enzyme located in mitochondrias. But if the effect of this enzyme is blocked by inhibiting drugs, alcohol or other amines, then the poisoning effect may be enhanced. Amines can influence the physiology by altering the vascular system (e.g., hypertensive crises, migraine, and inflammation), stimulating the secretion of gastric acid and affecting the nervous system. The toxicity threshold may vary depending on the detoxification mechanisms of different individuals.

Another potential hazard would be the reaction of polyamines, like spermine and spermidine, with nitrite under acidic conditions to form volatile nitrosamines (Nakamura *et al.* 1979). Preventive measures to reduce the production of active amines should exist in the application of good manufacturing practices, with special attention to the hygiene chain and the quality of the raw material. In fact, cadaverine, putrescine, histamine and tyramine combined can provide a useful index to evaluate meat freshness. Biogenic amines are not usually found in meat, but their hygienic quality is a critical factor. The use of starter cultures may also have a beneficial effect in reducing the formation of amines.

TABLE 6.1.
RELEVANT AMINES AND POLYAMINES, EXPRESSED AS MG/100 G AND WITH THEIR RESPECTIVE PRECURSOR AMINO ACIDS, IN DIFFERENT DRY-CURED MEAT PRODUCTS
(Adapted from Maijala *et al.* 1995, Hernández-Jover *et al.* 1997, Demeyer *et al.* 2000)

Amine	Formed from	Northern sausages	Mediterranean sausages	Dry-cured ham
Histamine	Histidine	0.1-0.2	0.1-3.6	0-0.1
Cadaverine	Lysine	0-0.1	0.3-0.6	0-30
Spermine	Methionine	0.4-3.0	2.0-3.0	3.0-4.0
Spermidine	"	0.1-0.4	0.5	5.0-6.0
Putrescine	Ornithine	0.1-2.8	2.0-10.0	0-1.7
Phenylethylamine	Phenylalanine	0.1-0.5	2.2-3.5	0-0.8
Tryptamine	Tryptophane	1.4-1.8	2.4-6.0	Non detected
Tyramine	Tyrosine	1.7-7.0	14.0-16.0	0-5.0

REFERENCES

ACTON, A.C. 1977. The chemistry of dry sausages. Proc. Annual Reciprocal Meat Conf. *30*, 49–62.

ASTIASARÁN, I., VILLANUEVA, R. and BELLO, J. 1990. Analysis of proteolysis and protein insolubility during the manufacture of some varieties of dry sausage. Meat Sci. *28*, 111–117.

BALDINI, P. *et al.* 2000. Dry sausages ripening: Influence of thermohygrometric conditions on microbiological, chemical and physico-chemical characteristics. Food Res. Int. *33*, 161–170.

BUSCAILHON, S., MONIN, G., CORNET, M. and BOUSSET, J. 1994. Time-related changes in nitrogen fractions and free amino acids of lean tissue of French dry-cured ham. Meat Sci. *37*, 449–456.

DEMASI, T.W., WARDLAW, F.B., DICK, R.L. and ACTON, J.C. 1990. Nonprotein nitrogen (NPN) and free amino acid contents of dry, fermented and nonfermented sausages. Meat Sci. *27*, 1–12.

DEMEYER, D.I. 1992. Meat fermentation as an integrated process. *In* New Technologies for Meat and Meat Products, J.M. Smulders, F. Toldrá, J. Flores and M. Prieto (eds.), pp. 21–36. Audet, Nijmegen, The Netherlands.

DEMEYER, D. *et al.* 2000. Control of bioflavor and safety in fermented sausages: first results of a European project. Food Research Int. *33*, 171–180.

DIERICK, N., VANDEHERCKHOVE, P. and DEMEYER, D. 1974. Changes in nonprotein nitrogen compounds during dry sausage ripening. J. Food Sci. *39*, 301–304.

FADDA, S., SANZ, Y., VIGNOLO, G., ARISTOY, M-C., OLIVER, G. and TOLDRÁ, F. 1999a. Hydrolysis of pork muscle sarcoplasmic proteins by *Lactobacillus curvatus* and *Lactobacillus sake*. Appl. Environ. Microbiol. *65*, 578–584.

FADDA, S., SANZ, Y., VIGNOLO, G., ARISTOY, M-C., OLIVER, G. and TOLDRÁ, F. 1999b. Characterization of muscle sarcoplasmic and myofibrillar protein hydrolysis caused by *Lactobacillus plantarum*. Appl. Environ. Microbiol. *65*, 3540–3546.

FLORES, M., SANZ, Y., SPANIER, A.M., ARISTOY, M.C. and TOLDRÁ, F. 1998. Contribution of muscle and microbial aminopeptidases to flavor development in dry-cured meat products. *In* Food Flavors: Generation, Analysis and Process Influence, Contis *et al.* (eds.), pp. 547–557. Elsevier Science B.V., Amsterdam, The Netherlands.

GARCÍA-GARRIDO, J.A., QUILES, R., TAPIADOR, J. and LUQUE, M.D. 2000. Activity of cathepsin B, D, H and L in Spanish dry-cured ham of normal and defective texture. Meat Sci. *56*, 1–6.

GOLL, D.E., OTSUKA, Y., NAGAINIS, P.D., SHANNON, J.D. and MUGURAMA, M. 1993. Role of muscle proteinases in maintenance of muscle integrity and mass. J. Food Biochem. *7*, 137–177.

HERNÁNDEZ-JOVER, T., IZQUIERDO, M., VECIANA, M.T., MARINÉ, A. and VIDAL-CAROU, M.C. 1997. Biogenic amine and polyamine contents in meat and meat products. J. Agric. Food Chem. *45*, 2098–2102.

KLEMENT, J.T., CASSENS, R.G. and FENNEMA, O.W. 1973. The association of protein solubility with physical properties in a fermented sausage. J. Food Sci. *38*, 1128–1131.

KLEMENT, J.T., CASSENS, R.G. and FENNEMA, O.W. 1974. The effect of bacterial fermentation on protein solubility in a sausage model system. J. Food Sci. *39*, 833–835.

KOOHMARAIE, M. 1994. Muscle proteinases and meat ageing. Meat Sci. *36*, 93–104.

LÜCKE, F.K. 1985. Fermented sausages. *In* Microbiology of Fermented Foods, B.J.B. Wood (ed.), Vol. 2, pp. 41–83. Elsevier Applied Science, London, U.K.

MAIJALA, R., EEROLA, S., LIEVONEN, S., HILL, P. and HIRVI, T. 1995. Formation of biogenic amines during ripening of dry sausages as affected by starter culture and thawing time of raw materials. J. Food Sci. *60*, 1187–1190.

MARTÍN, L., CÓRDOBA, J.J., ANTEQUERA, T., TIMÓN, M.L. and VENTANAS, J. 1998. Effects of salt and temperature on proteolysis during ripening of Iberian ham. Meat Sci. *49*, 145–153.

MOLINA, I. and TOLDRÁ, F. 1992. Detection of proteolytic activity in microorganisms isolated from dry-cured ham. J. Food Sci. *57*, 1308–1310.

MOLLY, K., DEMEYER, D., JOHANSSON, G., RAEMAEKERS, M., GHISTELINK, M. and GEENEN, I. 1997. The importance of meat enzymes in ripening and flavour generation in dry fermented sausages. First results of an European project. Food Chem. *59*, 539–545.

MONIN, G., MARINOVA, P., TALMANT, A., MARTÍN, J.F., CORNET, M., LANORE, D. and GRASSO, F. 1997. Chemical and structural changes in dry-cured hams (Bayonne hams) during processing and effects of the dehairing technique. Meat Sci. *47*, 29–47.

NAKAMURA, M., WADA, Y., AZUAYA, H. and KAWABATA, T. 1979. Polyamine content in fresh and processed pork. J. Food Sci. *44*, 515–517, 523.

ORDOÑEZ, J.A., HIERRO, E.M., BRUNA, J.M. and DE LA HOZ, L. 1999. Changes in the components of dry-fermented sausages during ripening. Crit. Rev. Food Sci. Nutr. *39*, 329–367.

PAROLARI, G., VIRGILI, R. and SCHIVAZZAPPA, C. 1994. Relationship between cathepsin B activity and compositional parameters in dry-cured hams of normal and defective texture. Meat Sci. *38*, 117–122.

RICO, E., TOLDRÁ, F. and FLORES, J. 1991. Effect of dry-curing process parameters on pork muscle cathepsins B, H and L activities. Z. Lebensm. Unter. Forsch. *193*, 541–544.

RODRIGUEZ-NUÑEZ, E., ARISTOY, M-C. and TOLDRÁ, F. 1995. Peptide generation in the processing of dry-cured ham. Food Chem. *53*, 187–190.

ROSELL, C.M. and TOLDRÁ, F. 1996. Effect of curing agents on m-calpain activity throughout the curing process. Z. Lebensm. Unters. Forchs. *203*, 320–325.

RUIZ, J., GARCÍA, C., DÍAZ, M.C., CAVA, R., TEJEDA, J.F. and VENTANAS, J. 1999. Dry cured Iberian ham non-volatile components as affected by the length of the curing process. Food Res. Int. *32*, 643–651.

SANTOS, N.N., SANTOS, R.C., SANZ, Y., BOLUMAR, T., ARISTOY, M-C. and TOLDRÁ, F. 2001. Hydrolysis of pork muscle sarcoplasmic proteins by *Debaryomyces hansenii*. Int. J. Food Microbiol. *68*, 199–206.

SANZ, B., SELGAS, D., PAREJO, I. and ORDÓÑEZ. J.A. 1988. Characteristics of lactobacilli isolated from dry fermented sausages. Int. J. Food Microbiol. *6*, 199–205.

SANZ, Y., FLORES, J., TOLDRÁ, F. and FERIA, A. 1997. Effect of preripening on microbial and chemical changes in dry fermented sausages. Food Microb. *14*, 575–582.

SANZ, Y., FADDA, S., VIGNOLO, G., ARISTOY, M-C., OLIVER, G. and TOLDRÁ, F. 1999a. Hydrolytic action of *Lactobacillus casei* CRL 705 on pork muscle sarcoplasmic and myofibrillar proteins. J. Agric. Food Chem. *47*, 3441–3448.

SANZ, Y., FADDA, S., VIGNOLO, G., ARISTOY, M-C., OLIVER, G. and TOLDRÁ, F. 1999b. Hydrolysis of muscle myofibrillar proteins by *Lactobacillus curvatus* and *Lactobacillus sake*. Int. J. Food Microbiol. *53*, 115–125.

SANZ, Y. and TOLDRÁ, F. 1997. Aminopeptidase activities from *Lactobacillus sake* in models of curing ingredients and processing conditions for dry sausages. J. Food Sci. *62*, 1211–1213, 1234.

SCHIVAZZAPPA, C., SACCANI, G., VIRGILI, R. and BORDINI, S. 1995. Changes in free amino acids during dry-curing of typical Italian raw ham. Ind. Conserve *70*, 377–385.

SENTANDREU, M.A., STOEVA, S., ARISTOY, M-C., LAIB, K., VOELTER, W. and TOLDRÁ, F. 2002. Identification of taste related peptides in Spanish Serrano dry-cured hams. J. Food Sci. (Accepted).

SENTANDREU, M.A. and TOLDRÁ, F. 2001. Dipeptidylpeptidase activities along the processing of Serrano dry-cured ham. Eur. Food Res. Technol. *213*, 83–87.

STRAUB, B.W., KICHERER, M., SCHILCHE, S.M. and HAMMES, W.P. 1995. The formation of biogenic amines by fermentation organisms. Z. Lebensm. Unters. Forsch. *201*, 79–82.

TARJÁN, V. and JÁNOSSY, G. 1978. The role of biogenic amines in foods. Die Nahrung *22*, 285–289.

TOLDRÁ, F. and ETHERINGTON, D.J. 1988. Examination of cathepsins B, D, H and L activities in dry cured hams. Meat Sci. *23*, 1–7.

TOLDRÁ, F., FLORES, J. and VOYLE, CH.A. 1990. Study of the white film developed on cut surfaces of vacuum-packed dry-cured ham. J. Food Sci. *55*, 1189–1191.

TOLDRÁ, F. 1992. The enzymology of dry-curing of meat products. *In* New Technologies for Meat and Meat Products, F.J.M. Smulders, F. Toldrá, J. Flores and M. Prieto (eds.), pp. 209–231. Audet, Nijmegen, The Netherlands.

TOLDRÁ, F., MIRALLES, M.-C. and FLORES, J. 1992. Protein extractability in dry-cured ham. Food Chem. *44*, 391–394.

TOLDRÁ, F., RICO, E. and FLORES, J. 1993. Cathepsin B, D, H and L activity in the processing of dry-cured ham. J. Sci. Food Agric. *62*, 157–161.

TOLDRÁ, F. 1998. Proteolysis and lipolysis in flavour development of dry-cured meat products. Meat Sci. *49*, S101–110.

TOLDRÁ, F. and FLORES, M. 1998. The role of muscle proteases and lipases in flavor development during the processing of dry-cured ham. CRC Crit. Rev. Food Sci. Nutr. *38*, 331–352.

TOLDRÁ, F., ARISTOY, M.-C. and FLORES, M. 2000. Contribution of muscle aminopeptidases to flavor development in dry-cured ham. Food Res. Int. *33*, 181–185.

TOLDRÁ, F., SANZ, Y. and FLORES, M. 2001. Fermentation Technology. *In* Meat Science and Applications, Y.H. Hui, H. Shorthose, O. Young, M. Koohmaraie and R.W. Rogers (eds.), pp. 537–561. Marcel Dekker Inc., New York.

TOLDRÁ, F. and VERPLAETSE, A. 1995. Endogenous enzyme activity and quality for raw product processing. *In* Composition of Meat in Relation to Processing, Nutritional and Sensory Quality, K. Lündstrom, I. Hansson and E. Winklund (eds.), pp. 41–55. ECCEAMST, Uppsala, Sweden.

TOLEDO, V.M., SELGAS, M.D., CASAS, M.C. ORDOÑEZ, J.A. and GARCÍA, M.L. 1997. Effect of selected mould strains on proteolysis in dry fermented sausages. Z. Lebensm. Unters. Forsch. A *204*, 385–390.

VANDEKERCKHOVE, P. 1977. Amines in dry fermented sausage. J. Food Sci. *42*, 283–285.

CHAPTER 7

CHARACTERIZATION OF LIPOLYSIS

LIPOLYSIS IN DRY-CURED HAM

As in the case of proteolysis, lipolysis also constitutes one of the most important group of biochemical reactions in the generation of flavor and/or flavor precursors during the processing of dry-cured ham. In fact, lipolysis has a high impact on quality for several reasons (Toldrá 1998): generation of free fatty acids with direct influence on flavor, generation of flavor precursors, such as free polyunsaturated fatty acids, that will act as substrates for further oxidative reactions to volatile compounds with specific aromas, contribution to fat texture by breakdown of triglycerides involved in the adipose tissue network and development of rancid aromas or yellowish colors in fat when there is an excess of lipolysis/oxidation.

In general, the most intense lipolytic changes have been observed during the initial five months of processing (Toldrá and Flores 1998). Lipolysis in dry-cured ham is attributed to the endogenous enzymatic systems in both the muscle and adipose tissues because the expected action of microbial lipases is reduced due to the low number of microorganisms inside the hams. Several research manuscripts published in the 1990s have focused on the muscle and adipose tissue lipolytic systems as a way to understand and optimize flavor development.

Action of Muscle and Adipose Tissue Lipases

The progress of lipolysis in dry-cured ham can vary depending on the type of product, the amount of endogenous lipolytic enzymes and specific process conditions. A description of main muscle lipolytic enzymes is given in Chap. 2. In general, lipolysis has the following stages (Fig. 7.1): initial lipids breakdown from the hydrolysis of major triglycerides by lipases and phospholipids by phospholipases; subsequent degradation of the generated di/monoglycerides and lysophospholipids by monoacylglycerol lipase and lysophospholipase activities, respectively; and generation of free fatty acids as final products of the lipolysis. As mentioned above, the action of microbial lipases has also been studied. Only an almost negligible lipase activity from microbial origin has been detected due to the low numbers of microorganisms usually found inside the hams.

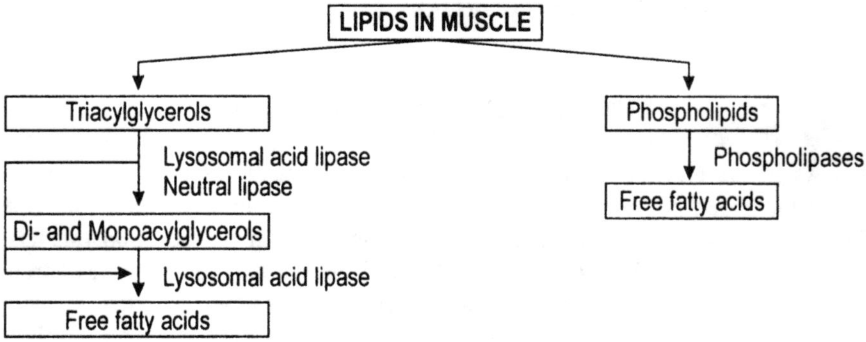

FIG. 7.1. GENERAL SCHEME SHOWING MAJOR STEPS IN MUSCLE LIPOLYSIS DURING THE PROCESSING OF DRY-CURED HAM

Muscle Lipases. Lysosomal acid lipase is active at acid pH (4.5–5.5), which is close to the levels found in postmortem meat, and neutral lipase is active at pH 7.0–7.5. Both lipases have a preference for primary ester bounds of triacylglycerols (Imanaka *et al.* 1984). Phospholipases A1 and A2 play an important role in the biochemical pathways involving phospholipids degradation. They catalyze the hydrolysis of the 1-acyl and 2-acyl ester, respectively, of sn-3-phosphoglycerides at the lipid/water interface (Yuan *et al.* 1990). Acid and neutral esterases, which are very stable and active, are able to hydrolyze short-chain fatty acids from tri-, di- and monoacylglycerols (Motilva *et al.* 1992). The evolution of muscle lipases along the processing of dry-cured ham is shown in Fig. 7.2, and, as can be observed, these enzymes show good stability along the full process.

Adipose Tissue Lipases. These enzymes are involved in lipolytical degradation of tri-, di-, and monoglycerides and subsequent generation of free long-chain fatty acids in adipose tissue (Fig. 7.3). There are several lipases that are active in the neutral/basic pH range and are located in adipose tissue (Toldrá 1992). Hormone-sensitive lipase, also known as neutral lipase due to its optimal pH at 7.0, generates free fatty acids with the hydrolysis of the ester bond at positions sn-1 and sn-3 in tri- and diacylglycerols (Belfrage *et al.* 1984). Monoacylglycerol lipase, also active at neutral pH, hydrolyzes monoacylglycerols with no positional specificity. Lipoprotein lipase is active at basic pH 8.5 and, although it plays a minor role, can hydrolize primary esters because unsaturated monoacylglycerols hydrolyze at a faster rate than saturated compounds (Yuan *et al.* 1990). All these enzymes show activity up until the end

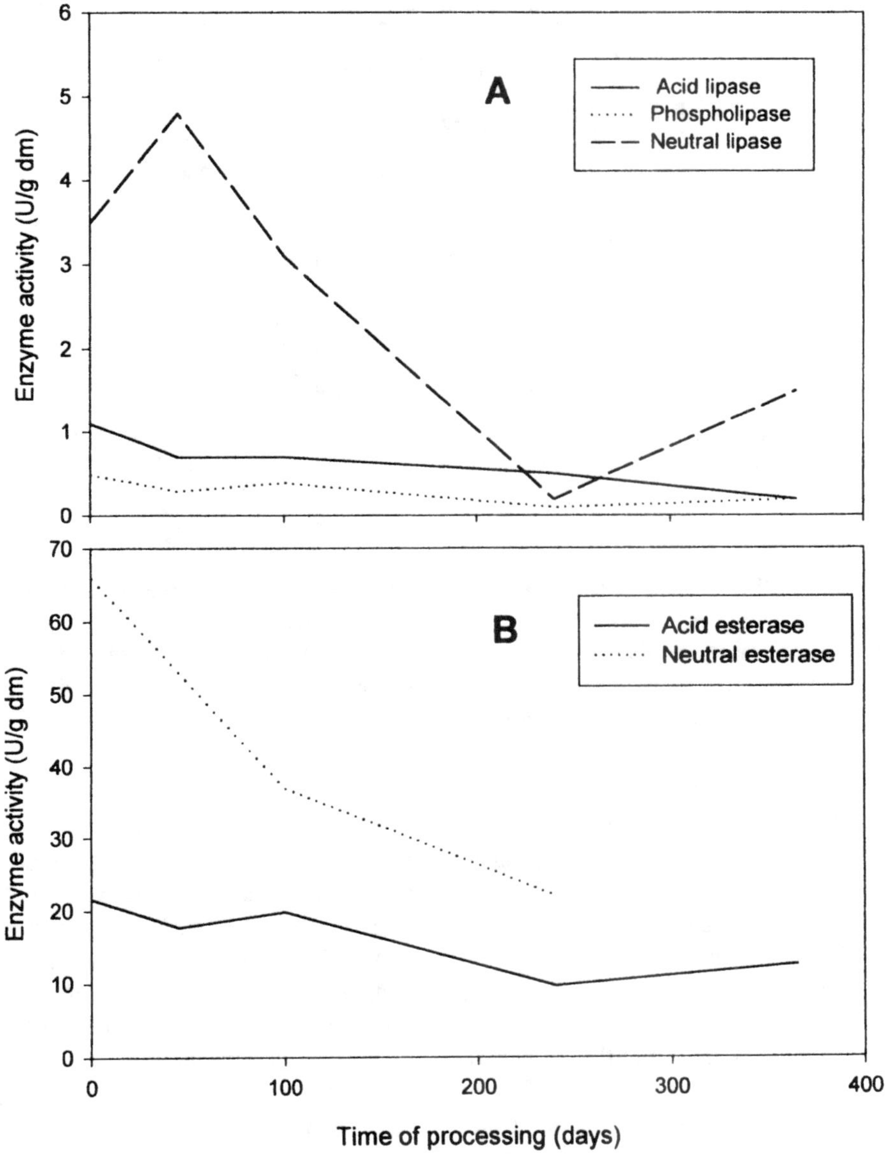

FIG. 7.2. EVOLUTION OF MUSCLE ENZYMES ACTIVITY IN THE MUSCLE *BICEPS FEMORIS* DURING THE PROCESSING OF SERRANO TYPE DRY-CURED HAM. (A) ACID AND NEUTRAL LIPASES AND ACID PHOSPHOLIPASE; (B) ACID AND NEUTRAL ESTERASES
(Toldrá 1996, unpublished data)

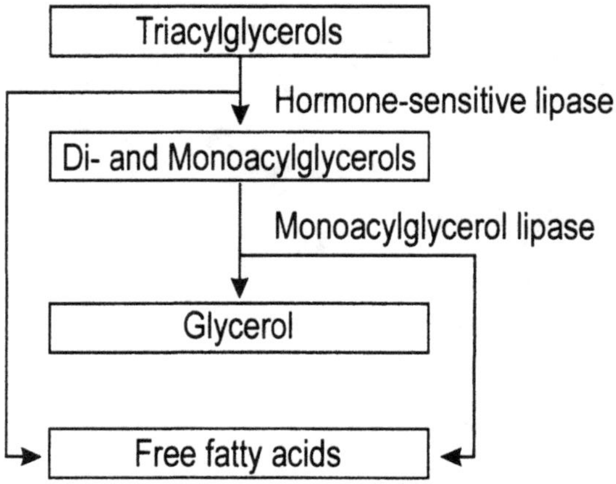

FIG. 7.3. GENERAL SCHEME SHOWING MAJOR STEPS IN LIPOLYSIS IN THE ADIPOSE TISSUE DURING THE PROCESSING OF DRY-CURED HAM

of post-salting stages, but only the neutral enzyme remains active during the ripening/drying period (Motilva *et al.* 1993b), as shown in Fig. 7.4. Acid and neutral esterases are also active and very stable in adipose tissue, although they generate very low amounts of short-chain fatty acids due to the poor availability of adequate substrate (Motilva *et al.* 1992).

Lipid Breakdown

The profile of lipid breakdown during the processing of dry-cured ham is shown in Fig. 7.5(A). The degradation of triglycerides is not as intense as expected. The generation of free fatty acids in muscle is correlated with the period of maximal phospholipid degradation (Flores *et al.* 1985; Motilva *et al.* 1993a; Buscailhon *et al.* 1994). The two main phospholipids, phosphatidylcholine and phosphatidylethanolamine, substantially decrease, and the composition of the generated free fatty acids (30.5% SFA, 25% MUFA and 45% PUFA) is very similar to the composition of the degraded phospholipids, that is, 35% SFA, 20% MUFA and 45% PUFA (Hernández *et al.* 1999).

A decrease in fatty acids from phospholipids is observed during the processing, especially in linoleic, arachidonic, oleic, palmitic and stearic acids (Martin *et al.* 1999). This decrease is more pronounced at early stages and may reach up to 66% of the total quantity of fatty acids of phospholipids. All of these facts also corroborate muscle phospholipases as the most important enzymes involved in muscle lipolysis. The amount of generated free fatty acids increases with aging time, with up to six months of processing being higher in the external

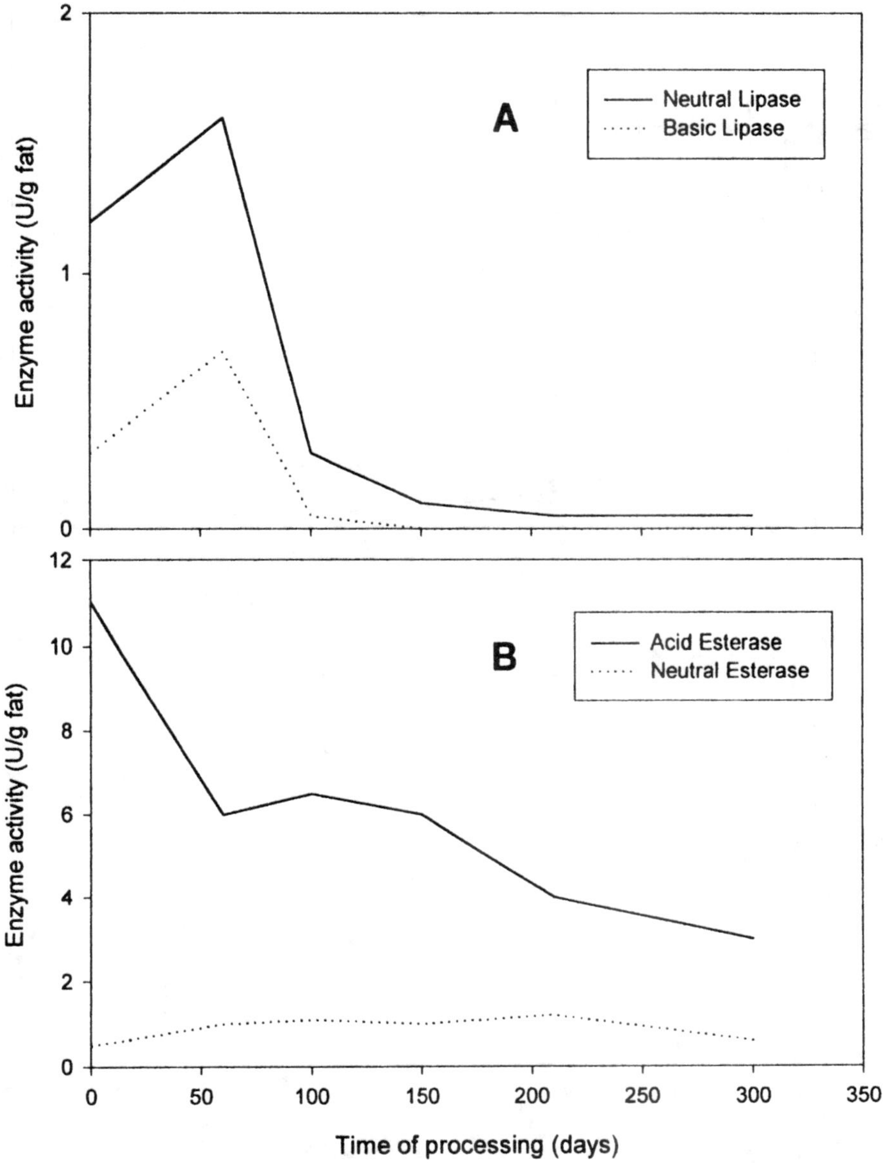

FIG. 7.4. EVOLUTION OF ENZYMES ACTIVITY IN THE ADIPOSE TISSUE DURING THE PROCESSING OF SERRANO TYPE DRY-CURED HAM. (A) NEUTRAL AND BASIC LIPASE. (B) ACID AND NEUTRAL ESTERASES
(Toldrá 1996, unpublished data)

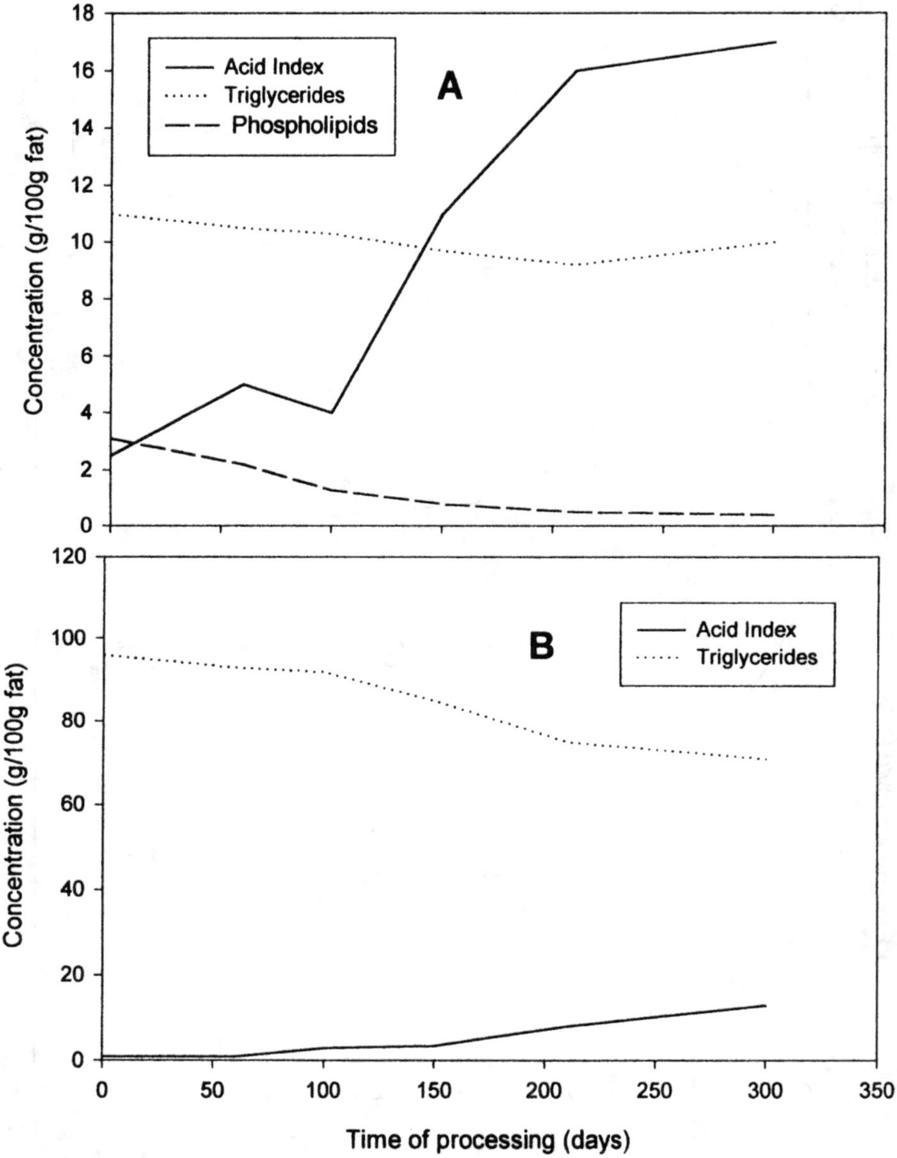

FIG. 7.5. PATTERNS OF LIPIDS BREAKDOWN AT DIFFERENT STAGES OF THE PROCESSING OF SERRANO DRY-CURED HAM. (A) MUSCLE BICEPS FEMORIS, (B) ADIPOSE TISSUE
(Toldrá 1996, unpublished data)

muscle (*Semimembranosus*), which contains more salt and is more dehydrated than the internal muscle (*Biceps femoris*). Because lipase activity is mainly influenced by pH, salt concentration and water activity (Toldrá and Motilva 1993), it appears that the observed lipid hydrolysis is favored by the same variables (salt increase and a_w reduction) that enhance enzyme activity *in vitro* (Vestergaard *et al.* 2000).

In the case of adipose tissue, triglycerides are forming the major part of this tissue (around 90%) and are mostly hydrolyzed by neutral lipase to di- and monoglycerides, as well as free fatty acids (Fig. 7.5(B)), especially and including up to six months of process (Motilva *et al.* 1993b). The amount of triglycerides decrease from about 90% down to 76% (Coutron-Gambotti and Gandemer 1999). There is a preferential hydrolysis of polyunsaturated fatty acids although some of them may not accumulate due to further oxidation during processing. Some triacylglycerols, rich in oleic and linoleic acids and liquid at 14–18C, are more hydrolyzed than others rich in saturated fatty acids, such as palmitic acid and solid at those temperatures (Coutron-Gambotti and Gandemer 1999). This means that the physical state of the triacylglycerols would increase the lipolysis rate by favoring the action of lipases at the water-oil interface.

Generation of Free Fatty Acids

The relative amounts of generated free fatty acids depend on the raw materials, especially the composition of the feed and type of process. The generation rate is high for up to 10 months of process. Afterwards, the profile remains asymptotic or even decreases as a consequence of further oxidative reactions. An example of the free fatty acids profile in muscle for each type of ham at the end of each respective process is shown in Fig. 7.6.

The rate of release of individual fatty acids decreases in the order linoleic > oleic > palmitic > stearic > arachidonic, which is very similar to the generation rate linoleic > oleic > linolenic > palmitic > stearic > arachidonic, observed during *in vitro* incubations of muscle lipases with pure phospholipids (Toldrá, unpublished results). Oleic, linoleic, stearic and palmitic acids appear as those accumulated in higher amounts not only because of their greater amounts in the initial lipid fraction but also because of their improved stability against oxidation. The amount of oleic acid is particularly high in Ibérico and Parma hams. On the other hand, the generation of short-chain free fatty acids is very low (Motilva *et al.* 1993a; Buscailhon *et al.* 1994).

The intensity of lipolysis in adipose tissue is also very high, particularly during the first stages of the process. Some of the fatty acids generated in greater amounts are oleic, palmitic, linoleic, stearic, palmitoleic and myristic acids (Fig. 7.7). Short-chain fatty acids are generated at very low levels (Motilva *et al.* 1993b; Buscailhon *et al.* 1994).

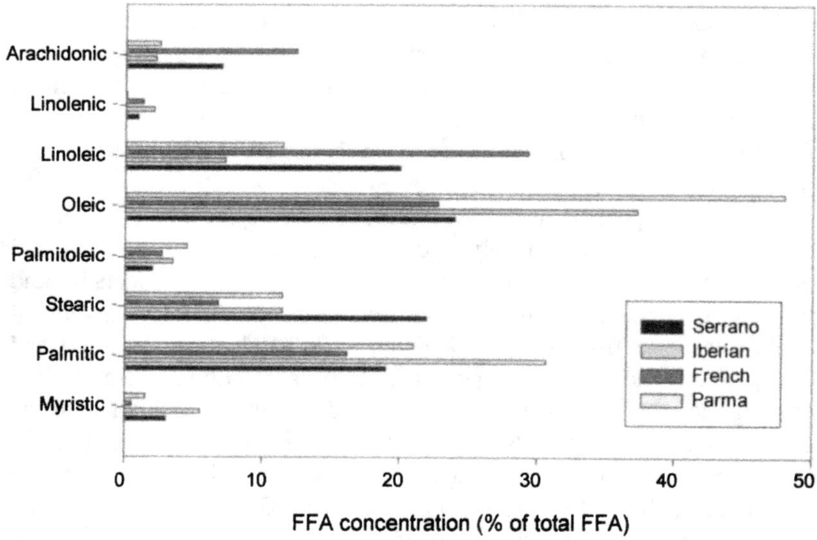

FIG. 7.6. MUSCLE ACCUMULATION OF FREE FATTY ACIDS IN SERRANO, IBERIAN, PARMA AND FRENCH DRY-CURED HAMS
(Adapted from Motilva *et al.* 1993a, Martín *et al.* 1999, Buscailhon *et al.* 1994 and Baldini *et al.* 1992)

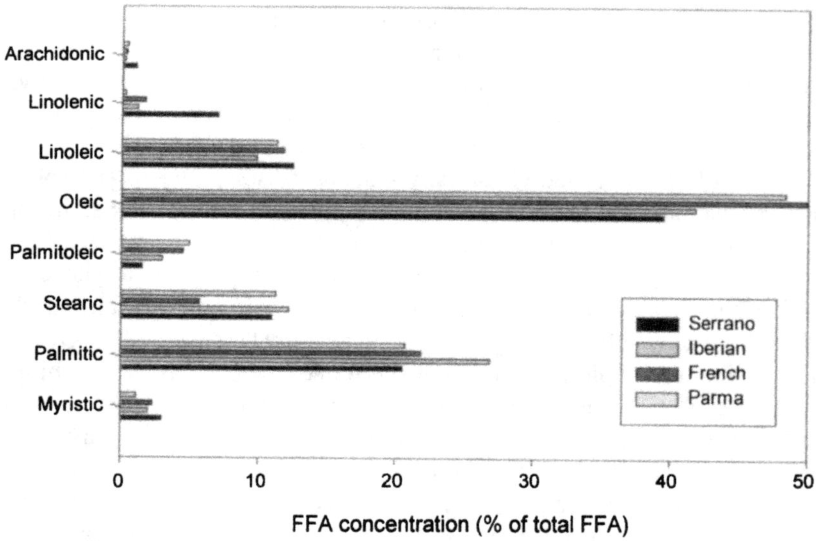

FIG. 7.7. ADIPOSE TISSUE ACCUMULATION OF FREE FATTY ACIDS IN SERRANO, IBERIAN, PARMA AND FRENCH DRY-CURED HAMS
(Adapted from Motilva *et al.* 1993b, Ruiz *et al.* 1998, Coutron-Gambotti and Gandemer 1999 and Baldini *et al.* 1992)

The analysis of the composition in fatty acids has been proposed as an effective method to determine the feeding background during the late fattening period (Ruiz *et al.* 1998). This is especially important in expensive products like Iberian ham, where a change of feed produces strong changes in the fatty acid profile. For instance, oleic acid is the major fatty acid present when pigs are fed acorns, and a higher percentage of saturated fatty acids can be found when pigs are fed commercial feeds. This affects the generated volatile compounds responsible of aroma resulting in a sensible decrease in quality.

However, adipose tissue, which is mainly composed of triglycerides, has a low fatty acid turnover rate and metabolic activity. The fattening period is exceeded by the average life for a triglyceride, and thus, the composition in fatty acid does not reflect the feed intake (Ruiz *et al.* 1998). Furthermore, lipid composition also varies depending on the place of sampling and from the internal to external layer (Flores *et al.* 1988). Other tissues such as liver have been proposed as better indexes of the feeding background (Ruiz *et al.* 1998).

LIPOLYSIS IN DRY-FERMENTED SAUSAGES

In modern fermentation processes, flavor development and consistent product quality have become as important as preservation. The increasing knowledge of both the endogenous and microbial lipolytic enzymes may be useful for accelerated ripening of dry-fermented sausages, modified generation of free fatty acids for flavor improvement and correction of some flavor defects.

As in the case of dry-cured ham, lipolysis also constitutes an important group of biochemical reactions in the generation of flavor and/or flavor precursors during the processing of dry-fermented sausages (Toldrá 1992). The degree of contribution of endogenous lipases and those of microbial origin, either naturally present in the product or added as starter cultures, will mainly depend on the raw materials, type of product and processing conditions (Demeyer 1992; Molly *et al.* 1997).

Action of Muscle and Microbial Lipases

The establishment of the different enzymes from different microbial groups is very difficult and complicated. In general, the amount of lipase produced is dependent on the environment, such as composition of the nitrogen, carbon and lipid sources (short-chain fatty acids stimulate its production while long chain fatty acids inhibit it), inorganic salts, cultivation temperature and availability of oxygen. Most bacterial lipases are stable at neutral or slightly acid pH. The pH dependence of activity and stability may vary according to culture conditions and the presence of adequate substrate, probably due to the adsorption to lipid-water interfaces. In this sense, a hydrophobic region should play an important role in catalysis in view of the hydrophobic nature of the interaction. Hydrolytic patterns may also change with temperature. So, the rate of hydrolysis of long-chain fatty acids increases with temperature.

Furthermore, there is a diversity in microbial flora and an evident overlapping with the endogenous enzyme system. It is necessary to elucidate the importance of bacterial and meat enzymes in lipolysis in order to be able to steer the process. Even though fatty acids are released in higher amounts when starters are added, there is a significant lipolysis in the absence of microbial starters (Montel *et al.* 1993; Hierro *et al.* 1997).

Mixtures of antibiotics and antimycotics have been used in sterile meat model systems to evaluate the relative importance of endogenous and microbial enzymes in lipolysis (Molly *et al.* 1996). The antibiotics prevented normal development of bacteria, but lipolysis was not affected in the presence of a lipolytic *Micrococcaceae*. The absence of glucose decreases carbonyl production irrespectively of the addition of antibiotics, and thus, carbonyl production appears to be independent of bacterial activity. The pattern of lipolysis is not affected by antibiotics. Thus, it was concluded that lipolysis is mainly brought about by muscle and fat tissue lipases (60 to 80% of total free fatty acids generated), with some variability depending on the batch and presence of specific strains (Molly *et al.* 1997). Lipases have preferences for long-chain fatty acids esterified at positions sn-1 and sn-3 of the triglyceride molecule, and phospholipases have preferences for phospholipids with a higher content in polyunsaturated fatty acids (Hierro *et al.* 1997). Bacteria would play a role in the formation of aldehydes from the carbohydrate fraction but not from the lipid fraction.

Useful species are selected as starter cultures, usually in mixtures of two or three and belonging to the genera *Lactobacillus*, *Pediococcus*, nonpathogenic *Staphylococcus* and *Kocuria* (Berdagué *et al.* 1993). *Micrococacceae* have a quite variable lipolytic activity that depends on the strain and type of substrate. In fact, large increases in acidity and very poor increases, depending on the strain, have been reported (Ordoñez *et al.* 1999). They have extra- and intracellular lipolytic enzymes. The extracellular enzymes become more important after 15–20 days of ripening. At this time, microorganisms decrease, acting on triglycerides and hydrolyzing them. In some cases, the lipolytic activity is too low. Other microorganisms with significant lipolytic activity are *Staphylococcus*. Several species that are commonly added as starters in French sausages have been studied (Montel *et al.* 1993). *S. warneri* gives the highest lipolytic activity followed by *S. saprophyticus*. *S. carnosus* and *S. xylosus* give poor and variable lipolytic activity.

Lactic acid bacteria also have some lipolytic activity, although the amount of activity is minor when compared to other microorganisms. Most of its lipolytic enzymes are intracellular. The production of these enzymes is correlated with the exponential growth phase and is stimulated by low glucose concentration. The maximal lipase production is observed at the optimal growth temperature and neutral pH (Papon and Talon 1988). The extracellular enzymes are especially active against di- and monoglycerides as well as triglycerides esterified to short-chain fatty acids. However, the lipolytic activity is much lower as the chain length increases and no activity is detected when fatty acids contain more than six carbon atoms (Sanz *et al.* 1998). Intracellular enzymes are almost unable to hydrolyze triglycerides. Most of the yeasts, such as *Candida*, *Cryptococcus* and *Trichosporum*, and molds isolated from fermented sausages are lipolytic and can contribute to flavor (Ordoñez *et al.* 1999).

Lipids Breakdown and Generation of Free Fatty Acids

The amount of free fatty acids increases along the process either in traditional or industrial dry-fermented sausages. Increases in the levels of free fatty acids may reach up to 2.5–5% of the total fatty acids, depending on the type of sausage and processing conditions. The majority of free fatty acids are released from triglycerides in the neutral fraction. The rate of release of individual fatty acids decreases in the order of linoleic > oleic > stearic > palmitic (Demeyer *et al.* 1974). Since there is a preponderance of unsaturated fatty acids in position sn-3 of triglycerides, this indicates some kind of specificity (Dainty and Blom 1995). However, there is a high specificity for release of polyunsaturated fatty acids from the polar lipid fraction (about 20% of the original concentration) in comparison to only 6% from triglycerides (Molly *et al.* 1997). The resulting increases in polyunsaturated fatty acids is 3.5 fold, 2.7 fold in monounsaturated fatty acids and 1.5 fold in saturated fatty acids. Oleic acid appears as the fatty acid generated in a higher relative amount, followed by palmitic, stearic and linoleic acids (Demeyer *et al.* 2000). Volatile fatty acids, such as acetic acid, also increase during the early stages of ripening. Free fatty acid content depends on raw materials and the length of the process. Mediterranean sausages have a substantial amount of lipolysis, and Norwegian sausages do too, although in this case, it is probably due to the raw materials (Demeyer *et al.* 2000). Phospholipids do not change during the fermentation, but they do experience a sensible decrease during ripening that is well correlated to free fatty acid generation (Navarro *et al.* 1997) as shown in Fig. 7.8.

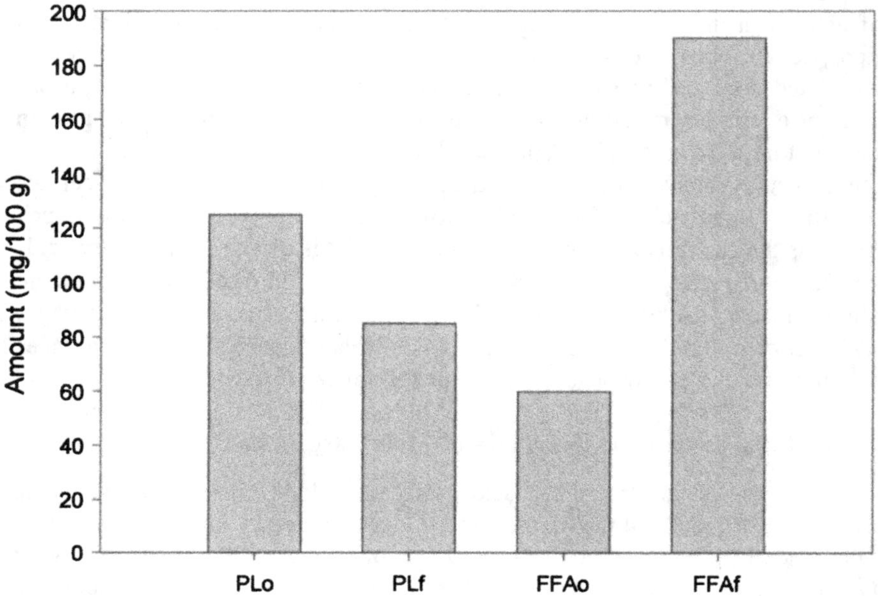

FIG. 7.8. VARIATION IN PHOSPHOLIPIDS (PL) AND FREE FATTY ACIDS (FFA) BETWEEN INITIAL (O) AND FINAL (F) PROCESSING OF DRY-FERMENTED SAUSAGES
(Adapted from Navarro *et al.* 1997)

OTHER ENZYMATIC REACTIONS

Nitrate Reductase Activity

The use of nitrate alone requires the presence of microorganisms with nitrate reductase activity like *Micrococacceae*. These strains reduce nitrate to nitrite, but an excessive acid pH must be prevented in order to avoid an early inhibition of the desired metabolic activity. Nitrate reductase, which transforms the nitrate present in the sausage to nitrite, is an intracellular enzyme that is formed at the cytoplasmic membrane. The enzyme is stable with salt and active at neutral pH, but its activity varies depending on the different strains (Jessen 1995). The addition of nitrate in the media stimulates the production of nitrate reductase activity. In some strains, like *S. carnosus* and *S. aureus*, activity increases up to fivefold while in other strains, like *S. xylosus* and *S. saprophyticus*, the effect is weak or even negligible (Talon *et al.* 1999).

Sometimes, nitrate may interact with carbohydrate fermentation or slow down the growth of lactic acid bacteria. The nitrite curing salt is necessary at the start of the fermentation for guaranteeing the microbial stability of the sausages and reducing the number of Enterobacteriaceae and psychrotrophs (Sanz *et al.* 1997). The amount of residual nitrites are usually very low in the final product. For instance, an initial 200 ppm level (with or without added starter culture) is reduced to 10–20 ppm after fermentation (Fig. 7.9) and less than 10 ppm after 30 days of ripening (Sanz *et al.* 1997).

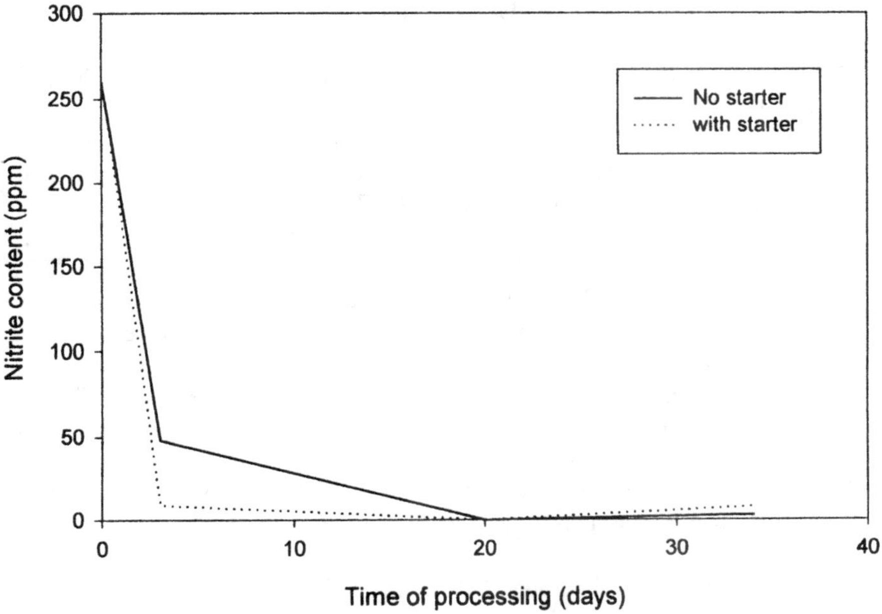

FIG. 7.9. EVOLUTION OF NITRITE CONCENTRATION DURING THE PROCESSING OF DRY-FERMENTED SAUSAGE WITH OR WITHOUT ADDED STARTER
(Adapted from Sanz *et al.* 1997)

Catalase Activity

This enzyme degrades the peroxides that are formed during the processing of the sausage, retarding the development of discoloration and rancidity and therefore stabilizing both the color and flavor of the final product. The catalase activity may be induced in the presence of glucose, but not by exogenous hydrogen peroxide. Catalase is inhibited by large amounts of salt, which effect is even more pronounced when combined with low pH values. The activity of catalase increases with cell growth to a maximum at the onset or during the stationary phase. In the case of dry-fermented sausages, catalase is formed during the ripening stage. The release of catalase in *Staphylococci* is not a common feature. For instance, *S. carnosus* have a high catalase activity in anaerobic conditions like those found in sausage processing. However, *S. warneri* has low catalase activity and, when this strain is present in sausages, it may result in rancidity (Talon *et al.* 1999).

REFERENCES

ACTON, A.C. 1977. The chemistry of dry sausages. Proc. Annual Reciprocal Meat Conf. *30*, 49–62.

BALDINI, P. *et al.* 2000. Dry sausages ripening: Influence of thermohygrometric conditions on microbiological, chemical and physico-chemical characteristics. Food Res. Int. *33*, 161–170.

BALDINI, P., BELLATTI, M., CAMORALI, G., PALMIA, F., PAROLARI, G., REVERBERI, M., PEZZANI, G., GUERRIERI, C., RACZYNSKI, R. and RIVALDI, P. 1992. Characterization of Italian raw ham by chemical, physical, microbiological and organoleptic parameters. Ind. Conserve *67*, 149–159.

BELFRAGE, P., FREDRIKSON, G., STRALFORS, P. and TORNQVIST, H. 1984. Adipose tissue lipases. *In* Lipases, B. Borgström and H.L. Brockman (eds.), pp. 365–416. Elsevier, Amsterdam.

BUSCAILHON, S., GANDEMER, G. and MONIN, G. 1994. Time-related changes in intramuscular lipids of French dry-cured ham. Meat Sci. *37*, 245–255.

CHASCO, J., BERIAIN, M.J. and BELLO, J. 1993. A study of changes in the fat content of some varieties of dry sausage during the curing process. Meat Sci. *34*, 191–204.

COUTRON-GAMBOTTI, C. and GANDEMER, G. 1999. Lipolysis and oxidation in subcutaneous adipose tissue during dry-cured ham processing. Food Chem. *64*, 95–101.

DAINTY, R. and BLOM, H. 1995. Flavour chemistry of fermented sausages. *In* Fermented Meats, G. Campbell-Platt and P.E. Cook (eds.), pp. 176–193. Blackie Academic & Professional, London.

DEMEYER, D.I., HOOZEE, J. and MESDOM, H. 1974. Specificity of lipolysis during dry sausage ripening. J. Food Sci. *39*, 293–296.

DEMEYER, D.I. 1992. Meat fermentation as an integrated process. *In* New Technologies for Meat and Meat Products, J.M. Smulders, F. Toldrá, J. Flores and M. Prieto (eds.), pp. 21–36. Audet, Nijmegen, The Netherlands.

DEMEYER, D. *et al.* 2000. Control of bioflavour and safety in fermented sausages: first results of a European project. Food Research Int. *33*, 171–180.

FLORES, J., BIRON, C., IZQUIERDO, L. and NIETO, P. 1988. Characterization of green hams from Iberian pigs by fast analysis of subcutaneous fat. Meat Sci. *23*, 253–262.

FLORES, J., NIETO, P., BERMELL, S. and MIRALLES, M.C. 1985. Changes in lipids during the slow and rapid maturation processes of dry cured ham and its relation with the quality. Rev. Agroquim. Tec. Aliment. *25*, 117–124.

HERNÁNDEZ, P., NAVARRO, J.L. and TOLDRÁ, F. 1999. Lipolytic and oxidative changes in two Spanish pork loin products: dry-cured loin and pickled-cured loin. Meat Sci. *51*, 123–128.

HIERRO, E., DE LA HOZ, L. and ORDOÑEZ, J.A. 1997. Contribution of microbial and meat endogenous enzymes to the lipolysis of dry fermented sausages. J. Agric. Food Chem. *45*, 2989–2995.

IMANAKA, T., AMANUMA-MUTO, K., OHKUMA, S. and TAKANO, T. 1981. Characterization of lysosomal acid lipase purified from rabbit liver. J. Biochem. *96*, 1089–1101.

JESSEN, B. 1995. Starter cultures for meat fermentation. *In* Fermented Meats, G. Campbell-Platt and P.E. Cook (eds.), pp. 130–159. Blackie Academic & Professional, London.

LÜCKE, F.K. 1985. Fermented sausages. *In* Microbiology of Fermented Foods, B.J.B. Wood (ed.), Vol. 2, pp. 41–83. Elsevier Applied Science, London.

MARTIN, L., CÓRDOBA, J.J., VENTANAS, J. and ANTEQUERA, T. 1999. Changes in intramuscular lipids during ripening of Iberian dry-cured ham. Meat Sci. *51*, 129–134.

MOLINA, I. and TOLDRÁ, F. 1992. Detection of proteolytic activity in microorganisms isolated from dry-cured ham. J. Food Sci. *57*, 1308–1310.

MOLLY, K., DEMEYER, D., CIVERA, T. and VERPLAETSE, A. 1996. Lipolysis in a Belgian sausage: Relative importance of endogenous and bacterial enzymes. Meat Sci. *43*, 235–244.

MOLLY, K., DEMEYER, D., JOHANSSON, G., RAEMAEKERS, M., GHISTELINK, M. and GEENEN, I. 1997. The importance of meat enzymes in ripening and flavour generation in dry fermented sausages. First results of an European project. Food Chem. *59*, 539–545.

MONTEL, M.C., TALON, R., BERDAGUÉ, J.L. and CANTONNET, M. 1993. Effects of starter cultures on the biochemical characteristics of French dry sausages. Meat Sci. *35*, 229–240.

MOTILVA, M.-J., TOLDRÁ, F. and FLORES, J. 1992. Assay of lipase and esterase activities in fresh pork meat and dry-cured ham. Z. Lebensm. Unters Forsch. *195*, 446–450.

MOTILVA, M.-J. and TOLDRÁ, F. 1993. Effect of curing agents and water activity on pork muscle and adipose subcutaneous tissue lipolytic activity. Z. Lebensm. Unters Forsch. *196*, 228–231.

MOTILVA, M.-J., TOLDRÁ, F., NIETO, P. and FLORES, J. 1993a. Muscle lipolysis phenomena in the processing of dry-cured ham. Food Chem. *48*, 121–125.

MOTILVA, M.-J., TOLDRÁ, F., ARISTOY, M.-C. and FLORES, J. 1993b. Subcutaneous adipose tissue lipolysis in the processing of dry-cured ham. J. Food Biochem. *16*, 323–335.

NAVARRO, J.L., NADAL, M.I., IZQUIERDO, L. and FLORES, J. 1997. Lipolysis in dry cured sausages as affected by processing conditions. Meat Sci. *45*, 161–168.

ORDOÑEZ, J.A., HIERRO, E.M., BRUNA, J.M. and DE LA HOZ, L. 1999. Changes in the components of dry-fermented sausages during ripening. Crit. Rev. Food Sci. Nutr. *39*, 329–367.

PAPON, M. and TALON, R. 1988. Factors affecting growth and lipase production by meat lactobacilli strains and Brochothrix thermosphacta. J. Appl. Bacteriol. *64*, 107.

RUIZ, J., CAVA, R., ANTEQUERA, T., MARTIN, L., VENTANAS, J. and LÓPEZ-BOTE, C. 1998. Prediction of the feeding background of Iberian pigs using the fatty acid profile of subcutaneous, muscle and hepatic fat. Meat Sci. *49*, 155–163.

SANZ, B., SELGAS, D., PAREJO, I. and ORDÓÑEZ. J.A. 1988. Characteristics of lactobacilli isolated from dry fermented sausages. Int. J. Food Microbiol. *6*, 199–205.

SANZ, Y., FLORES, J., TOLDRÁ, F. and FERIA, A. 1997. Effect of preripening on microbial and chemical changes in dry fermented sausages. Food Microbiol. *14*, 575–582.

TALON, R., WALTER, D., CHARTIER, S., BARRIÈRE, C. and MONTEL, M.C. 1999. Effect of nitrate and incubation conditions on the production of catalase and nitrate reductase by staphylococci. Int. J. Food Microbiol. *52*, 47–56.

TOLDRÁ, F. 1992. The enzymology of dry-curing of meat products. *In* New Technologies for Meat and Meat Products, F.J.M. Smulders, F. Toldrá, J. Flores and M. Prieto (eds.), pp. 209–231. Audet, Nijmegen, The Netherlands.

TOLDRÁ, F. 1998. Proteolysis and lipolysis in flavour development of dry-cured meat products. Meat Sci. *49*, S101–110.

TOLDRÁ, F. and FLORES, M. 1998. The role of muscle proteases and lipases in flavor development during the processing of dry-cured ham. CRC Crit. Rev. Food Sci. Nutr. *38*, 331–352.

TOLDRÁ, F., SANZ, Y. and FLORES, M. 2001. Meat Fermentation Technology. *In* Meat Science and Applications, Y.H. Hui, H. Shorthose, O. Young, M. Koohmaraie and R.W. Rogers (eds.), pp. 537–561. Marcel Dekker Inc., New York.

TOLDRÁ, F. and VERPLAETSE, A. 1995. Endogenous enzyme activity and quality for raw product processing. *In* Composition of Meat in Relation to Processing, Nutritional and Sensory Quality, K. Lündstrom, I. Hansson and E. Winklund (eds.), pp. 41–55. ECCEAMST, Uppsala, Sweden.

VESTERGAARD, C.S., SCHIVAZZAPPA, C. and VIRGILI, R. 2000. Lipolysis in dry-cured ham maturation. Meat Sci. *55*, 1–5.

YUAN, W., QUINN, D.M., SIGLER, P.B. and GELB, M.H. 1990. Kinetic and inhibition studies of phospholipase A2 with short chain substrates and inhibitors. Biochem. *29*, 6082–6094.

CHAPTER 8

FLAVOR DEVELOPMENT

Flavor generation constitutes a very complex process that involves numerous chemical and biochemical reactions and mainly affects proteins and lipids. The general scheme showing flavor generation during the processing of dry-cured ham is shown in Fig. 8.1. Proteolysis and lipolysis generate a great number of nonvolatile and volatile compounds with influence on taste and aroma. A similar approach, although more complex due to the involvement of microbial metabolism, is shown for dry-fermented sausages (Fig. 8.2).

The generation of aroma compounds is mainly determined by endogenous enzymes in hams, but in the case of fermented sausages, the metabolic properties of the microbial cultures, as described in Chap. 5, play an important role. So, proteolytic and lipolytic processes, as well as the development of secondary flora, have an influence on the final flavor profile. For instance, there is an important contribution of lactobacilli, streptococci and lactococci in dairy products such as cheese (Imhof and Bosset 1994).

GENERATION OF NONVOLATILE COMPOUNDS

Peptides and free amino acids are generated by proteolysis during the processing of dry-cured meat products, as described in Chap. 6. These compounds are known to contribute to meat taste during cooking (Spanier and Miller 1993) and/or aging and curing (Aristoy and Toldrá 1995). Some L-amino acids, such as alanine, serine, proline, glycine and hydroxyproline, contribute to sweet taste, while phenylalanine, tryptophan, arginine, methionine, valine, leucine and isoleucine contribute to bitterness (Katoh *et al.* 1989). Sourness is mainly provided by glutamic and aspartic acids, histidine and asparagine, and saltiness is provided by sodium salts of both glutamic and aspartic acids. Finally, umami taste, also known as the fifth taste, is associated with glutamic acid/monosodium glutamate.

The taste of the peptides is more complicated and depends on the nature of the amino acids, position, etc. The generation rate is more or less pronounced, depending on the type of product and length of the process. Several amino acids, such as glutamic acid, aspartic acid, methionine, isoleucine, leucine, and lysine, have been found to be correlated with the length of the drying process of Spanish Serrano ham and with both the cured and pork flavor (Flores *et al.* 1997a). Amounts of peptides and free amino acids have been reported as very high in long-ripened products such as dry-cured ham (Aristoy and Toldrá 1991;

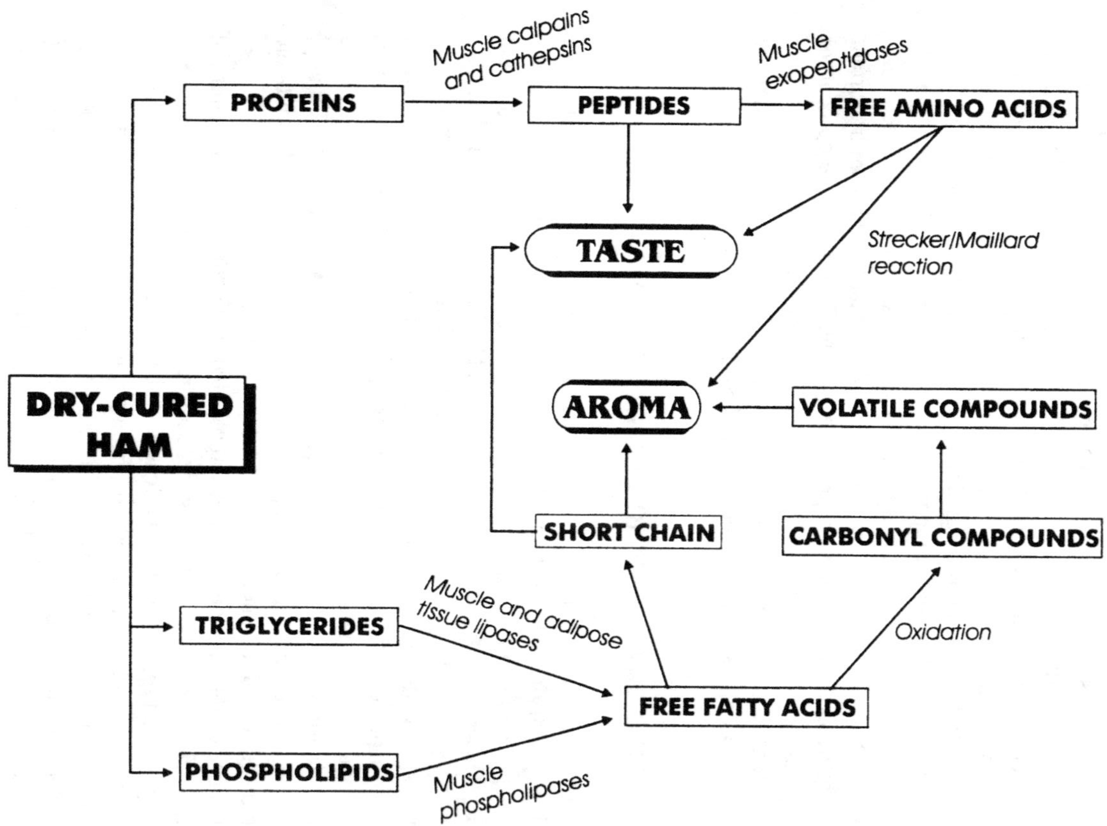

FIG. 8.1. GENERAL SCHEME SHOWING FLAVOR GENERATION DURING THE PROCESSING OF DRY-CURED HAM
(Adapted from Toldrá and Flores 1998)

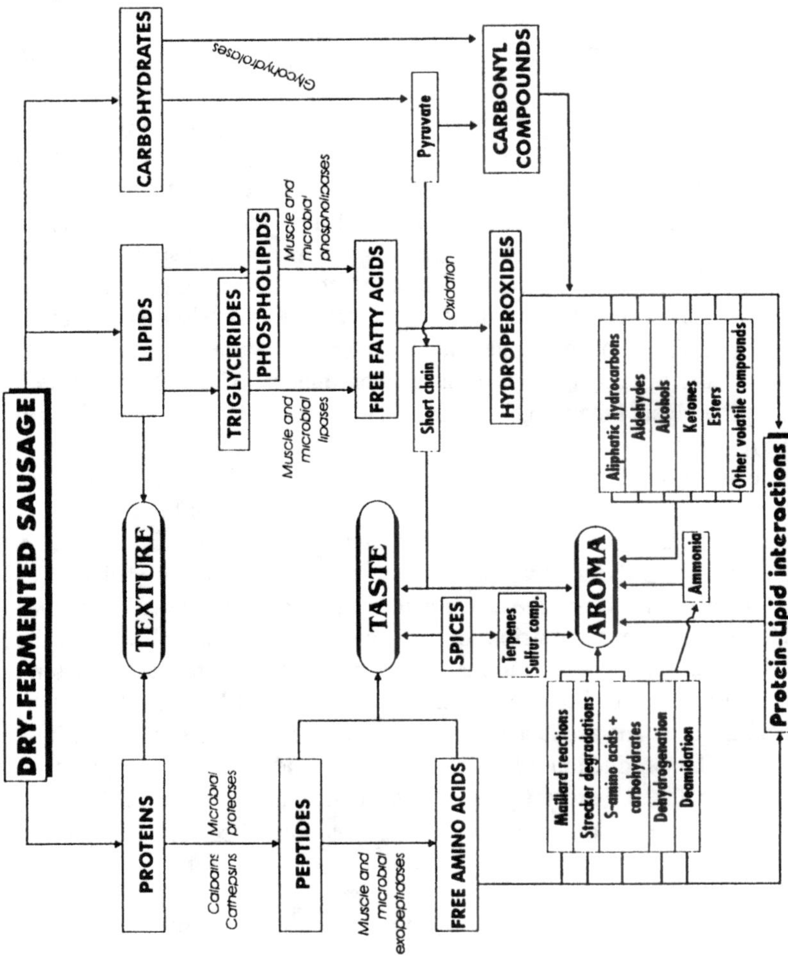

FIG. 8.2. GENERAL SCHEME SHOWING FLAVOR GENERATION DURING THE PROCESSING OF DRY-FERMENTED SAUSAGES
(Adapted from Toldrá and Flores 1998)

Toldrá *et al.* 1995). Some of the small peptides, those below 2500 Da and fractionated by the size of molecular mass, showed different tastes such as sour, salty, bitter. A fraction with ham flavor included some nucleotides and some compounds of protein-lipid interaction (Aristoy and Toldrá 1995). In other cases, like country-style ham (country-style ham is cooked before being eaten), the increases are not so important, but are still sufficient to provide a source of volatile compounds when the ham is heated, giving aged flavor (McCain *et al.* 1968).

Free amino acids and small peptides are taste-active and may exert a strong influence on the final flavor of the product. For instance, glutamic acid has been related to saltiness, and lysine and tyrosine have been related to an improvement in the aged taste of Parma ham (Careri *et al.* 1993). Furthermore, phenylalanine and isoleucine contribute positively, and tyrosine contributes negatively to the acid taste. However, only a small effect of these compounds on flavor development has been reported in French-type dry-cured ham (Buscailhon *et al.* 1994). Finally, it should be mentioned that some unpleasant tastes such as bitter-like or metal aftertaste may appear as a consequence of excessive protein hydrolysis (proteolysis index higher than 29–30%) (Careri *et al.* 1993; Virgili *et al.* 1995).

In the case of dry-fermented sausages, the amounts of generated peptides and free amino acids depend not only on the type and length of the process but also on the formulations, starter cultures and fermentation conditions. A noticeable increase in the generation of free amino acids is usually observed, although it mainly depends on the acid pH reached in the sausage (Flores *et al.* 1998a). However, the real contribution of proteolysis is mainly based on the sensory properties and concentrations of amino acids and peptides, but no concluding results have been published (Dainty and Blom 1995). Other compounds like lactic and acetic acids are generated by glycolysis and contribute to acid taste. The production of D-lactic acid is undesirable as it gives an unpleasant spicy taste. Free fatty acids also contribute to taste, although a relatively high concentration is needed for a perceptible effect on taste (Toldrá *et al.* 2001).

GENERATION OF VOLATILE COMPOUNDS

Most of the volatile compounds generated during the processing of dry-cured meat products are the result of chemical or enzymatic oxidation of unsaturated fatty acids and further interactions with proteins, peptides and free amino acids. A good number of volatile compounds have been reported in dry-cured hams (more than 200 summarized in Tables 8.1–8.3) and dry-fermented sausages (more than 150 summarized in Tables 8.4–8.6). Some of the most important are hydrocarbons (alkanes and methyl-branched alkanes), aldehydes, alcohols, ketones, free fatty acids, resulting from the hydrolysis of triglycerides and phospholipids (Motilva *et al.* 1992), β-lactones coming from the dehydration and cyclization of the β-hydroxyacids (Berdagué and García 1990), esters and other compounds such as benzene derivatives, amines and amides. The contribution of each compound to flavor depends on its characteristic aroma and the odor threshold. In general, alkanes are almost odorless. Aldehydes have green aromas, but in some cases they also have rancid aromas. Some ketones have buttery notes, and lactones give fruity aromas. Thus, the typical aroma of dry-cured meat products is mainly associated with a high number of volatile compounds that are generated during the processing through the following reactions.

Degradation of Free Amino Acids

There is an important contribution of free amino acids to the generation of volatile compounds, such as aldehydes, alcohols and ketones, that is very often found in dry-cured meat products and with a high incidence on the aroma. For instance, 2-methyl propanal, 2-methyl butanal, and 3-methyl butanal arise from Strecker degradation of the amino acids valine, isoleucine and leucine, respectively. Other compounds, such as dimethyl disulfide with an aromatic impact due to the low flavor threshold, are mainly formed from the sulfur containing amino acids (methionine, cysteine, and cystine) via Strecker degradations to thiols or through complex enzymatic reactions. For instance, two pathways for the degradation of methionine have been found in starter bacteria. The first pathway is by the enzyme cystathionine β-lyase which produces methanethiol. The second pathway has two steps: (1) there is transamination and the formation of 4-methylthio-2-oxobutyric acid, and (2) there is decarboxylation to methional, enzymatic conversion to methanethiol and then, to dimethyldisulphide (Smit *et al.* 2000). The end-products of amino acid metabolism are estimated as 6% of the total volatiles (Dainty and Blom 1995).

Reactions Between Amino Acids and Other Compounds

This constitutes another source of volatile compounds contributing to flavor development. Some pyrazines with a diversity of aromas (nutty, green, earthy, etc.) are formed through Maillard reactions between sugars and free amino acids. The amount of generated pyrazines in dry-cured meat products is generally low because the temperature used during dry-curing, even during the fermentation, is not as high as in cooking. Some pyridines may be formed by a reaction of aldehydes with amino compounds. Furans are also formed through the reaction of sulfur containing amino acids with carbohydrates and have a burned rubber odor that turns into a pleasant roasted note at high dilutions (Shahidi *et al.* 1986).

TABLE 8.1.
COMPILATION OF MAIN ALIPHATIC HYDROCARBONS AND ALDEHYDES IDENTIFIED IN SPANISH SERRANO (SER) AND IBERIAN (IBE), ITALIAN PARMA (PAR), FRENCH BAYONNE (BAY) AND CORSICAN (COR) AND AMERICAN COUNTRY-STYLE (C/S) DRY-CURED HAMS

Compound	Aroma	Ser[a]	Ibe[b]	Par[c]	Bay[d]	Cor[e]	C/S[f]
Aliphatic Hydrocarbons							
Pentane	Alkane			X	X		
Hexane	Green-mold	X		X	X	X	
Heptane	Alkane	X	X	X	X		
Octane	Alkane	X	X	X	X		
Nonane	Crackers	X			X		
Decane	Alkane	X	X		X		
Undecane	"		X		X		
Branched alkanes	"		X		X		
Aldehydes							
2- methyl butanal	Roasted cocoa		X	X		X	
3-methyl butanal	Cheesy-green	X	X	X	X	X	
Methanal	Pungent				X		
Butanal	"		X	X	X		X
Pentanal	"	X	X	X	X	X	X
Hexanal	Green-grassy	X	X	X	X	X	
Heptanal	Fatty	X	X	X			X
Octanal	Green-fresh	X	X	X	X	X	
Nonanal	Green	X	X		X	X	
Decanal	Soapy		X	X			X
Undecanal	Pungent		X		X		X
Dodecanal	Lily		X		X		X
Alk-2-enals		X	X	X			X
Alk-2,4-di-enals		X	X	X			X

[a]Flores *et al.* 1997, [b]Ruiz *et al.* 1998, García *et al.* 1991, [c]Barbieri *et al.* 1992, Bolzoni *et al.* 1996, [d]Berdagué *et al.* 1991, Buascailhon and Monin 1994, [e]Sabio *et al.* 1998, [f]Lillard and Ayres 1969

TABLE 8.2.
COMPILATION OF MAIN ALCOHOLS AND KETONES IDENTIFIED IN SPANISH SERRANO (SER) AND IBERIAN (IBE), ITALIAN PARMA (PAR), FRENCH BAYONNE (BAY) AND CORSICAN (COR) AND AMERICAN COUNTRY-STYLE (C/S) DRY-CURED HAMS

Compound	Aroma	Ser[a]	Ibe[b]	Par[c]	Bay[d]	Cor[e]	C/S[f]
Alcohols							
1-penten-3-ol	Onion-toasted	X		X	X	X	
2- methyl 1-butanol	Wine		X		X		
3-methyl 1-butanol	Penetrating green	X	X	X	X	X	
3-methyl 2-hexanol	Potato-wheat	X				X	
1-butoxy 2-propanol				X			
Methanol			X				X
Ethanol	Sweet	X			X	X	
Propanol	Alcoholic	X	X	X			X
Butanol	Medicinal	X	X	X	X		
Pentanol	Balsamic	X	X	X	X	X	
Hexanol	Resinous	X	X	X	X		X
Heptanol	Chemical		X	X	X	X	X
Octanol	Mushroom	X	X	X	X		X
Decanol			X				
Ketones							
Propanone		X			X		
2-butanone	Buttery	X	X	X	X		X
2,3-butanedione	"	X	X		X	X	
2,3-pentanedione	Creamy		X		X		
2-pentanone diacetyl	Ethereal						X
Pentanone	"			X	X	X	
2-hexanone	Floral-apple	X			X	X	
2,6-hexanedione				X	X	X	X
6-me-5-heptan-2-one	Citrus-candy	X			X		
2-heptanone	Soapy	X	X		X	X	
2-octanone	"		X		X	X	
2-nonanone	Hot milk		X			X	

[a]Flores *et al.* 1997, [b]Ruiz *et al.* 1998, García *et al.* 1991, [c]Barbieri *et al.* 1992, Bolzoni *et al.* 1996, [d]Berdagué *et al.* 1991, Buascailhon and Monin 1994, [e]Sabio *et al.* 1998, [f]Lillard and Ayres 1969

TABLE 8.3.
COMPILATION OF MAIN ESTERS AND AROMATIC HYDROCARBONS IDENTIFIED IN SPANISH SERRANO (SER) AND IBERIAN (IBE), ITALIAN PARMA (PAR), FRENCH BAYONNE (BAY) AND CORSICAN (COR) AND AMERICAN COUNTRY-STYLE (C/S) DRY-CURED HAMS

Compound	Aroma	Ser[a]	Ibe[b]	Par[c]	Bay[d]	Cor[e]	C/S[f]
Esters							
Methyl acetate		X					X
Ethyl acetate	Pineapple				X		X
Butyl acetate	Pear				X	X	
Methyl,2-me-propanoate	Floral	X					X
Ethyl butanoate	Apple	X		X			
Methyl butanoate	Sweet-caramel	X		X			
Ethyl 2-me-butanoate	Fruity-strawberry	X	X	X	X		
Methyl hexanoate	Pineapple	X		X			X
Ethyl hexanoate	Apple peel	X	X	X	X	X	
Methyl octanoate	Orange			X			X
Methyl decanoate	Wine			X			X
Methyl hexadecanoate	Waxy				X		
Aromatic hydrocarbons							
Styrene	Medicinal	X		X	X		
o-xylene	Sweet-fruit candy	X	X	X	X		
p-/m- xylene	Smoked-phenolic	X	X	X	X		
Nitrogen compounds							
1-H-pyrrole	Meaty	X				X	
Methylpyrazine	Nutty	X		X	X		
2,6-dimethylpyrazine	Toasted nuts	X	X	X	X	X	
Furans							
2-ethylfuran				X	X		
2-pentylfuran	Ham-like	X	X	X	X		
2,5-dimethylfuran	Sulphury-fishy	X			X		
Sulfur compounds							
Dimethyldisulfide	Dirty socks	X		X	X		

[a]Flores *et al.* 1997, [b]Ruiz *et al.* 1998, García *et al.* 1991, [c]Barbieri *et al.* 1992, Bolzoni *et al.* 1996, [d]Berdagué *et al.* 1991, Buascailhon and Monin 1994, [e]Sabio *et al.* 1998, [f]Lillard and Ayres 1969

TABLE 8.4.
COMPILATION OF MAIN ALIPHATIC HYDROCARBONS, ACIDS AND ALDEHYDES IDENTIFIED IN SPANISH (S), FRENCH (F), ITALIAN (I) AND DANISH (D) DRY-FERMENTED SAUSAGES

Compound	Aroma	S[a]	F[b]	I[c]	D[d]
Aliphatic Hydrocarbons					
Pentane	Alkane		X		
Hexane	Green-mold		X		
Heptane	Alkane		X		
Octane	"		X		
Nonane	Crackers				
Acids					
Acetic acid	Sour		X	X	X
Butanoic acid	Rancid		X	X	X
3-methyl butanoic acid	Sweaty			X	X
2-methyl butanoic acid	"				X
2-methyl pentanoic acid			X	X	X
Hexanoic acid	Sweaty			X	X
Aldehydes					
2-methyl butanal	Roasted cocoa		X		X
3-methyl butanal	Cheesy-green	X	X		X
Butanal	Pungent				X
Pentanal	"	X	X		X
Hexanal	Green-grassy	X	X	X	X
Heptanal	Fatty				X
Octanal	Green-fresh			X	X
Nonanal	Green	X		X	X
Decanal	Soapy			X	X
Undecanal	Pungent				X

[a]Edwards *et al.* 1999, [b]Berdagué *et al.* 1993, [c]Berger *et al.* 1990, [d]Stanhke 1999

TABLE 8.5.
COMPILATION OF MAIN ALCOHOLS AND KETONES IDENTIFIED IN SPANISH (S), FRENCH (F), ITALIAN (I) AND DANISH (D) DRY-FERMENTED SAUSAGES

Compound	Aroma	S[a]	F[b]	I[c]	D[d]
Alcohols					
1-penten-3-ol	Onion-toasted		X		X
2- methyl 1-butanol	Wine	X			X
3-methyl 1-butanol	Penetrating green	X	X		X
3-methyl 2-butanol	Potato-wheat			X	X
2-butoxy ethanol	Dark toast-meaty		X		
Ethanol	Sweet				X
2-Propanol	Alcoholic		X		X
Butanol	Medicinal	X	X		X
Pentanol	Balsamic	X		X	X
Hexanol	Resinous	X			X
Heptanol	Chemical	X			
Octanol	Mushroom	X			
Ketones					
2-Propanone					X
2-butanone	Buttery				X
2,3-butanedione	"	X	X		X
2,3-pentanedione	Creamy	X			
2-pentanone	Ethereal		X	X	X
2-hexanone	Floral-apple		X		X
6-me-5-heptan-2-one	Citrus-candy	X			
2-heptanone	Soapy	X	X	X	X
2-octanone	"	X	X	X	
2-nonanone	Hot milk	X		X	X

[a]Edwards *et al.* 1999, [b]Berdagué *et al.* 1993, [c]Berger *et al.* 1990, [d]Stanhke 1999

TABLE 8.6.
COMPILATION OF MAIN ESTERS, AROMATIC HYDROCARBONS, TERPENES, FURANES AND SULFUR COMPOUNDS IDENTIFIED IN SPANISH (S), FRENCH (F), ITALIAN (I) AND DANISH (D) DRY-FERMENTED SAUSAGES

Compound	Aroma	S[a]	F[b]	I[c]	D[d]
Esters					
Methyl acetate					X
Ethyl acetate	Pineapple	X	X		X
Methyl,2-me-propanoate	Floral	X			
Ethyl butanoate	Apple	X			
Ethyl 2-me-butanoate	Fruity-strawberry	X			X
Ethyl hexanoate	Apple peel	X			
3-Methyl butylacetate					X
2-Methyl butylacetate					X
Aromatic hydrocarbons					
o-xylene	Sweet-fruit candy		X		
p-/m- xylene	Smoked-phenolic		X		
Terpenes					
Various	Pine oil	X		X	X
Furans					
2-methylfuran					X
2-pentylfuran	Ham-like		X		X
Sulfur compounds					
Dimethyldisulfide	Dirty socks		X		X

[a]Edwards *et al.* 1999, [b]Berdagué *et al.* 1993, [c]Berger *et al.* 1990, [d]Stanhke 1999

Oxidation

A high amount of free fatty acids are generated as a consequence of the lipolysis, as described in Chap. 7. Then, unsaturated fatty acids act as precursors of further oxidative reactions producing many of the volatile compounds responsible for the final aroma. The generation of free fatty acids, susceptible to oxidation, constitutes a key stage in flavor generation. A small amount of oxidation is needed to get the characteristic aroma in dry-cured meats, although an excess of oxidation may lead to off-flavors, rancidity and yellow colors. As oxidative rancidity reduces palatability scores, alternatives for retarding rancidity, such as removal of heavy metal ions from the salt or the addition of antioxidants, would have a desirable effect on acceptance of these hams. In fact, the generation of the characteristic aroma of dry-cured meat products is in agreement with the beginning of lipid oxidation (Buscailhon *et al.* 1993).

Oxidation has three definite stages. The first stage is initiation, where a free radical is formed. This reaction is catalyzed by light, moisture, heat, metallic cations and some oxidative enzymes (like peroxydases and cyclooxygenases). Propagation is the second stage, where the free radicals react with oxygen to form peroxide radicals. They react with double bonds forming hydroxyperoxides (primary oxidation product). The breakdown of hydroperoxides, which are very unstable, produces many types of secondary oxidation products by a free radical mechanism and propagates oxidation reactions because of the acceleration of the rate of lipid oxidation. The third stage is termination, where radicals react with each other and become inactivated.

Hydroperoxides are flavorless, but the secondary oxidation products contribute to flavor (Lillard 1978). The most important volatile compounds formed by oxidation of unsaturated fatty acids are aliphatic hydrocarbons, alcohols, aldehydes and ketones.

Aliphatic hydrocarbons result from the autooxidation of the lipids (Loury 1972) but do not contribute significantly to flavor.

Alcohols are mainly originated by oxidative decomposition of certain lipids. For instance, 1-propanol and 1-butanol may be derived from myristoleic acid, 1-pentanol from linoleic acid and 1-octanol from oleic acid. Alcohols have a relatively high odor threshold.

Aldehydes, resulting from free fatty acid oxidation, can react with other components to produce flavor chemicals. Hexanal is a typical aldehyde resulting from linoleic acid oxidation. These compounds may contribute to flavor even in trace amounts due to their low odor thresholds. Aldehydes may undergo further oxidation into shorter chain aldehydes. Enals and dienals appear during heating via autooxidation of the appropriate polyunsaturated fatty acids.

Ketones are produced either through β-keto acid decarboxylation or through fatty acid β-oxidation (Berdagué *et al.* 1991). For instance, 2-heptanone is an oxidation product of linoleic acid. About 60% of the total volatile compounds found in dry sausages result from lipid oxidation (Berdagué *et al.* 1993).

The esters are derived from the interaction of free carboxylic fatty acids and various alcohols generated by lipid oxidation in the intramuscular tissue (Shahidi *et al.* 1986). The aged odor of Italian type dry-cured ham is positively affected by esters, aromatic hydrocarbons and cyclic nitrogen compounds (Careri *et al.* 1993). In the case of French type dry-cured ham (Buscailhon *et al.* 1994), the relation of some aldehydes with the aroma of fresh-cured pork and ketones with the pleasant aroma of dry-cured ham is similar to that found in the Spanish dry-cured ham (Flores *et al.* 1997).

The esters have been found in greater amounts and are characteristic in the aroma of Italian dry-cured ham where nitrate is not used (Barbieri *et al.* 1992). Thus, the lower concentration of esters found in Spanish and French hams is probably due to the inhibitory effect of nitrate/nitrite on lipid oxidation. Something similar happens in sausages when TBA and carbonyl levels are low and do not increase during ripening.

In the case of sausages, oxidation mainly affects linoleic and oleic acids. There is a rapid buildup of peroxides during the first days of fermentation and then a rapid decrease during ripening because these compounds are quite unstable. TBA increases more markedly in products like Spanish chorizo than in French saucisson or Italian salami (Chasco *et al.* 1993). Volatile aldehydes, mainly hexanal are produced. Nitrites act against lipid oxidation through different mechanisms: binding heme and preventing release of catalytic iron, binding heme and nonheme iron inhibiting catalysis and stabilizing olefinic lipids against oxidation. Spices, like paprika and garlic, and smoke compounds, such as phenols, exert antioxidant properties.

Glycolysis

Lactic acid formation is stimulated in the presence of red pepper, probably due to its manganese content that may be involved in the Embden-Meyerhof pathway. Acetic acid is another acid produced from hexoses and amino acids, and its concentration depends on the type of bacteria inoculated in the sausage. The production of lactic acid must be controlled since an excess of acid results in an unpleasant sour taste that restricts consumer acceptance. Other compounds with 2-4 carbon atoms, like ethanol, 3-hydroxy-2-butanone, 2,3-butanodione and 2,3-butanediol may be produced (Edwards *et al.* 1999).

Other Compounds

The addition of spices is a common practice in the manufacture of dry-fermented sausages (Chap. 4). These aromatic plant substances can be used in the natural form, or as extracts, to give a characteristic flavor, and sometimes color as well, to the sausage. For instance, paprika contains carotenoids, as typical pigments of its color, capsantin, β-carotene, etc. and imparts a characteristic color in addition to the flavor properties. Garlic contributes with several volatiles, including sulphur containing compounds. Up to 27 have been identified, allicin being the main component of the volatile fraction (Ordoñez *et al.* 1999). Pepper contains piperine, responsible for its hot flavor, piperine isomers, piperidine, piperonylaldehyde, etc., and its terpene hydrocarbon content is about 90% (Ordoñez *et al.* 1999). Garlic and pepper also contain some antioxidative compounds that protect against autooxidation (Dainty and Blom 1995).

Feeds also constitute a source of various volatile compounds. This is the case of xylene isomers, found in plants, or chloride compounds from pesticide residues.

SENSORY CHARACTERISTICS

The description of the dry-cured ham flavor is complex, and often the flavor terms are subjective depending on the origin of the product. The development of a lexicon requires precise descriptors. Descriptive analysis are used by trained panels to identify and quantify the attributes of meat products. Once the lexicon is developed, the most representative attributes are selected for further analysis (Flores *et al.* 1998). Examples of typical descriptors for dry-cured hams are as follows (Flores *et al.* 1997c): aromatics associated with lipid products (fat complex), boar meat/hormone-like (boar taint), free fatty acids (barnyard), cooked pork meat (pork), tastes in the tongue associated with citric acid (sour), sodium ions (salty), caffeine (bitter) and mouthfeel associated with glutamic acid (mouthfilling). Other usual descriptors are: aroma associated with nutty/almonds (nutty), dill and vinegar (pickling spice), extremely oxidized fat (rancid), color homogeneity of transversal cuts (homogeneity), intramuscular fat (marbling) and presence of tyrosine crystals on transversal cut (crystals).

Processing time has a great influence on aroma and taste traits of ham. For instance, Iberian ham ripened for a longer time had higher values in flavor strength, cured flavor and aftertaste (Ruiz *et al.* 1998b). In the case of Serrano hams, the longer process gave higher barnyard, sour and salty taste and was characterized by high contents of 2-propanol, 2-propanone and heptane (Flores *et al.* 1998b). The short process was characterized by aldehydes, branched aldehydes, alcohol and dimethyl disulfide (fresh-cured pork flavor). Some amino acids like glutamic acid, aspartic acid, methionine, isoleucine, leucine and lysine were also related to the length of the process (Flores *et al.* 1997a). In Parma hams, the length of process was determined by 3-methyl butanal, ethyl esters and alcohols like propanol, 1-butoxy-2-propanol and 2-butanol that increased at 12 months (Bolzoni *et al.* 1996). In French hams, compounds present were 2-pentanone, 1-pentanol, ethanol, ethyl acetate, 1-pentene-3-ol and pentanal. Nonane increased up to 179 days, while methyl alcohols and a branched alkane decreased after that time (Buscailhon *et al.* 1993).

Differences in volatile compounds are found among the different types of hams, mainly due to the raw materials and the technology used. The rearing conditions, such as a free-range system based on acorn and pasture versus confinement on a commercial feed, affects sensory attributes related to lipid content and fatty acids composition. The hams from pigs raised on a free-range system gave higher scores in oiliness, brightness of the lean, marbling and aroma and flavor traits (Cava *et al.* 2000).

The comparison of dry-cured hams from different crossbreeds and sex revealed that hams from females and Duroc-sired pigs gave the best overall quality. Hams from Belgian Landrace-sired pigs gave poor quality and rancid aromas (Armero *et al.* 1999).

The length of the process constitutes an important factor of variation in quality attributes. It is evident that as the process time increases, a higher amount of volatile compounds are detected. So, Iberian and Corsican hams, characterized by longer processing times, higher intramuscular fat content and higher temperatures during drying, contain a higher amount of methyl ketones, aldehydes, branched aldehydes and alcohols. The accumulation of these compounds in longer processed Iberian hams is in agreement with higher scores for positive odor and flavor traits and lower scores for rancidity (Ruiz *et al.* 1999).

The volatile content in the rest of hams is lower, especially in those of shorter processing time like country-style hams. Bayonne and Corsican hams may be rich in terpenes if pepper is rubbed onto the surface. The location of the sample is also important. The external muscle *Semimembranosus* tends to be harder, drier and more fibrous than the internal *Biceps femoris* (Cava *et al.* 2000). The esters are formed in greater amounts in Parma hams because the process is different, and nitrate is not added with the curing salt.

RELATION BETWEEN SENSORY ANALYSIS AND FLAVOR

Many studies have been performed to correlate flavor compounds and sensory analysis, most of them using multivariate statistical methods of factor analysis. Some relations have been reported between nonvolatile components and product flavor. In the case of volatile compounds, some may even be dominant in absolute amounts, their contribution to flavor may be reduced due to particular odor characteristics, high odor threshold levels, concentration in the product, solubility in water or fat and temperature.

Dry-Cured Ham

The relationship between the sensory response and fatty acid concentration appears to depend on the phospholipids with which the fatty acid is associated. For instance, flavor characteristics of ground beef, as determined by a descriptive panel, were significantly correlated with total phospholipids and fatty acid content of the individual phospholipids (Larick and Turner 1990). Some amino acids are also related to flavor. So, lysine and tyrosine are associated with the higher quality taste of aged hams, and asparagine is associated with poor quality. Glutamic acid contributes to salty taste, and phenylalanine and isoleucine contribute to sour taste (Careri *et al.* 1993). However, the results with French-type ham indicate little effect of amino acids on development of the aroma of cured meat (Buscailhon *et al.* 1994).

Ketones, esters, aromatic hydrocarbons and pyrazines are correlated with a pleasant aroma of ham. On the other hand, hexanal, 3-methyl butanal and dimethyl disulfide are related with short drying processes (Flores *et al.* 1998). Methyl-branched aldehydes, secondary alcohols, methyl ketones, ethyl esters and dimethyl trisulfide are correlated with nutty, cheesy and salty descriptors (Hinrichsen and Pedersen 1995). This is very close to the correlation of short-chain methyl-branched aldehydes, esters and alcohols found with aged flavor of Parma ham (Careri *et al.* 1993; Bolzoni *et al.* 1996).

Dry-Fermented Sausages

When pH drop is pronounced, sausages contain high amounts of lactic acid, low amounts of proteolysis products and some aldehydes, ketones, alcohols and esters. No marked curing aroma is appreciated, there is only a predominant sour taste. When pH drop is milder, aldehydes and ketones, furans, sulfur compounds, pyrazines and amines are generated (Toldrá *et al.* 2001). Italian Milano salami is preferred when it is medium aged. Better preference is observed for salamis with a higher pH and lower amount of lactic acid. However, preference decreases when the product is ripened too much with a high salt/moisture ratio (Casiraghi *et al.* 1996). *Lactobacilli* and *P. pentosaceus* are mainly involved in pH reduction. *Staphylococcus* and *Kocuria* contribute to free fatty acid generation, nitrate reduction and microbial metabolism related with free fatty acids (Montel *et al.* 1993). The generation of the volatile compounds depends on the specific starter cultures used. Some compounds, like diacetyl, acetoin or butanediol with butter-yogurt aroma, are generated during the fermentation. Three types of descriptors are influenced by the type of starter (Berdagué *et al.* 1993): odor of butter correlated with acetoin, diacetyl, 1,3-butanediol and 2,3-butanediol, which are related to the addition of *S. saprophyticus* and *S. warneri;* curing odor correlated with 2-pentanone, 2-hexanone and 2-heptanone and related to the addition of *S. carnosus* + *P.*

acidilactici, *S. carnosus* + *L. sakei*, *S. carnosus* + *P. pentosaceus*; and lower rancidity related with *S. saprophyticus*. Nitrite plays a role in the development of specific flavor characteristics in dry-fermented sausages, contributing to a stronger and a more typical flavor (Noel *et al.* 1990). The addition of nitrate to nonfermented sausages contributes to a higher intensity in aroma and taste (Sanz *et al.* 1998).

Italian and Spanish sausages contain the following main compounds: ketones and aldehydes (from lipid autooxidation), terpenes (from spices), some short-chain free fatty acids (from lipolysis), alcohols and some nitrogen-derived volatile compounds (from proteolysis). In some products like Spanish fuet (small diameter, mild conditions and short processing time), most of the volatiles, like ketones, aldehydes and alcohols, are produced by lipid autooxidation. There is a lack of esters or sulphur compounds in these types of products. The longer ripening time of Spanish salchichón results in a higher microbial activity, and thus in the additional generation of esters. Some terpenes like α-pinene, β-pinene, β-mircene, e-carene and limonene are also present from pepper (García-Regueiro *et al.* 1998). In the case of Spanish chorizo, the microbial activity is also high and results in the production of esters. For instance, *Staphylococcus xylosus* LTH produces a considerable amount of methyl and ethyl esters when used as starter (García-Regueiro *et al.* 1998). Some sulphur compounds are derived from the garlic that is characteristically added to this product. The spices and condiments, such as pepper, paprika, mustard, oregano, rosemary, garlic and onion, have a high impact on the aroma.

REFERENCES

ARISTOY, M-C. and TOLDRÁ, F. 1991. Deproteinization techniques for HPLC amino acid analysis in fresh pork muscle and dry-cured ham. J. Agric. Food Chem.*39*, 1792–1795.

ARISTOY, M-C. and TOLDRÁ, F. 1995. Isolation of flavor peptides from raw pork meat and dry-cured ham. *In* Food Flavors: Generation, Analysis and Process Influence, G. Charalambous (ed.), pp. 1323–1344. Elsevier Science Publishers B.V., Amsterdam, The Netherlands.

ARMERO, E., FLORES, M., TOLDRÁ, F., BARBOSA, J-A., OLIVET, J., PLA, M. and BASELGA, M. 1999. Effects of pig sire types and sex on carcass traits, meat quality and sensory quality of dry-cured ham. J. Sci. Food Agric. *79*, 1147–1154.

BERDAGUÉ, J.L., DENOYER, C., LE QUERÉ and SEMON, E. 1991. Volatile compounds of dry-cured ham. J. Agric. Food Chem. *39*, 1257–1261.

BERDAGUÉ, J.L., MONTEIL, P., MONTEL, M.C. and TALON, R. 1993. Effects of starter cultures on the formation of flavour compounds in dry sausage. Meat Sci. *35*, 275–287.

BERGER, R.G., MACKU, C., GERMAN, J.B. and SHIBAMOTO, T. 1990. Isolation and identification of dry salami volatiles. J. Food Sci. *55*, 1239–1242.

BOLZONI, L., BARBIERI, G. and VIRGILI, R. 1996. Changes in volatile compounds of Parma ham during maturation. Meat Sci. *43*, 301–310.

BUSCAILHON, S., BERDAGUÉ, J.L. and MONIN, G. 1993. Time-related changes in volatile compounds of lean tissue during processing of French dry-cured ham. J. Sci. Food Agric. *63*, 69–75.

BUSCAILHON, S., BERDAGUÉ, J.L., BOUSSET, J., CORNET, M., GANDEMER, G., TOURAILLE, C. and MONIN, G. 1994. Relations between compositional traits and sensory qualities of French dry-cured ham. Meat Sci. *37*, 229–243.

BUSCAILHON, S. and MONIN, G. 1994. Determinisme des qualités sensorielles du jambon sec. Viandes Prod. Carnés *15*, 23–34.

CARERI, M., MANGIA, A., BARBIERI, G., BOLZONI, L., VIRGILI, R. and PAROLARI, G. 1993. Sensory property relationships to chemical data of Italian type dry-cured ham. J. Food Sci. *58*, 968–972.

CASIRAGHI, E., POMPEI, C., DELLAGLIO, S., PAROLARI, G. and VIRGILI, R. 1996. Quality attributes of Milano salami, an Italian dry-cured sausage. J. Agric. Food Chem. *44*, 1248–1252.

DAINTY, R. and BLOM, H. 1995. Flavour chemistry of fermented sausages. *In* Fermented Meats, G. Campbell-Platt and P.E. Cook (eds.), pp. 176–193. Blackie Academic & Professional, London.

EDWARDS, R.A., ORDOÑEZ, J.A., DAINTY, R.H., HIERRO, E.M. and DE LA HOZ, L. 1999. Characterization of the headspace volatile compounds of selected Spanish dry fermented sausages. Food Chem. *64*, 461–465.

FLORES, M., ARISTOY, M-C., SPANIER, A.M. and TOLDRÁ, F. 1997a. Non-volatile components effects on quality of Serrano dry-cured ham as related to processing time. J. Food Sci. *62*, 1235–1239.

FLORES, M., GRIMM, C.C., TOLDRÁ, F. and SPANIER, A.M. 1997b. Correlations of sensory and volatile compounds of Spanish Serrano dry-cured ham as a function of two processing times. J. Agric. Food Chem. *45*, 2178–2186.

FLORES, M., INGRAM, D.A., BETT, K.L., TOLDRÁ, F. and SPANIER, A.M. 1997c. Sensory characteristics of Spanish "Serrano" dry-cured ham. J. Sensory Studies *12*, 169–179.

FLORES, M., SANZ, Y., SPANIER, A.M., ARISTOY, M.C. and TOLDRÁ, F. 1998a. Contribution of muscle and microbial aminopeptidases to flavor development in dry-cured meat products. *In* Food Flavor: Generation, Analysis and Process Influence, Contis *et al.* (eds.), pp. 547–557. Elsevier Science B.V., Amsterdam, The Netherlands.

FLORES, M., SPANIER, A.M. and TOLDRÁ, F. 1998b. Flavour analysis of dry-cured ham. *In* Flavor of Meat, Meat Products and Seafoods, pp. 320–341. F. Shahidi (ed.), Blackie Academic & Professional, London.

GARCÍA, C., BERDAGUÉ, J.L., ANTEQUERA, T., LÓPEZ-BOTE, C., CÓRDOBA, J.J. and VENTANAS, J. 1991. Volatile compounds of dry cured Iberian ham. Food Chem. *41*, 23–32.

GARCÍA-REGUEIRO, J.A., RIUS, M.A., HORTÓS, M., HUGAS, M. and DÍAZ, I. 1998. Influence of starter cultures on the volatile flavor profile of a Spanish fermented sausage. 44th Intern. Congr. Meat Science and Technology, Vol. II, pp. 802–803. Estrategias Alimentarias, Madrid, Spain.

HINRICHSEN, L.L. and PEDERSEN, S.B. 1995. Relationship among flavor, volatile compounds, chemical changes and microflora in Italian-type dry-cured ham during processing. J. Agric. Food Chem. *43*, 2932–2940.

IMHOF, R. and BOSSET, J.O. 1994. Relationships between microorganisms and formation of aroma compounds in fermented dairy products. Z. Lebensm. Unters. Forsch. *198*, 267–276.

LARICK, D.K. and TURNER, B.E. 1990. Flavor characteristics of forage and grain fed beef as influenced by phospholipid and fatty acid compositional differences. J. Food Sci. *55*, 312–317, 368.

LILLARD, D.A. and AYRES, J.C. 1969. Flavor compounds in Country cured hams. Food Technol. *23*, 117–120.

LILLARD, D.A. 1978. Chemical changes involved in the oxidation of lipids in foods. *In* Lipids as a Source of Flavor. M.K. Supran (ed.), pp. 68–80. ACS Symp. Series 75, Washington, D.C.

McCAIN, G.R., BLUMER, T.N., CRAIG, H.B. and STEEL, R.G. 1968. Free amino acids in ham muscle during successive aging periods and their relation to flavor. J. Food Sci. *33*, 142–146.

MONTEL, M.C., TALON, R., BERDAGUÉ, J.L. and CANTONNET, M. 1993. Effects of starter cultures on the biochemical characteristics of French dry sausages. Meat Sci. *35*, 229–240.

RUIZ, J., CAVA, R., VENTANAS, J. and JENSEN, T. 1998a. Headspace solid phase microextraction for the analysis of volatiles in a meat product: Dry-cured Iberian ham. J. Agric. Food Chem. *46*, 4688–4694.

RUIZ, J., VENTANAS, J., CAVA, R., TIMÓN, M.L. and GARCÍA, C. 1998b. Sensory characteristics of Iberian ham: influence of processing time and slice location. Food Res. Int. *31*, 53–58.

RUIZ, J., VENTANAS, J., CAVA, R., ANDRÉS, A. and GARCÍA, C. 1999. Volatile compounds of dry-cured Iberian ham as affected by the length of the curing process. Meat Sci. *52*, 19–27.

SABIO, E., VIDAL-ARAGON, M.C., BERNALTE, M.J. and GATA, J.L. 1998. Volatile compounds present in six types of dry-cured ham from south European countries. Food Chem. *61*, 493–503.

SANZ, Y., VILA, R., TOLDRÁ, F. and FLORES, J. 1998. Effect of nitrate and nitrite curing salts on microbial changes and sensory quality of non-fermented sausages. Int. J. Food Microbiol. *42*, 213–217.

SMIT, G., VERHEUL, A., KRANENBOURG, R., AYAD, E., SIEZEN, R. and ENGELS, W. 2000. Cheese flavour development by enzymatic conversions of peptides and amino acids. Food Res. Int. *23*, 153–160.

SPANIER, A.M. and MILLER, J.A. 1993. Role of proteins and peptides in meat flavor. *In* Food Flavour and Safety. Molecular Analysis and Design, A.M. Spanier, H. Okai, and M. Tamura (eds.), pp. 78–97. ACS Symposium Series 258, American Chemical Society, Washington, D.C.

STANHKE, L.H. 1999. Volatiles produced by Staphylococcus xylosus and Staphylococcus carnosus during growth in sausage minces. Part I. Collection and identification. Lebensm.-Wiss. U-Technol. *32*, 357–364.

TOLDRÁ, F., FLORES, M. and ARISTOY, M-C. 1995. Enzyme generation of free amino acids and its nutritional significance in processed pork meats. *In* Food Flavors: Generation, Analysis and Process Influence, G. Charalambous (ed.), pp. 1303–1322. Elsevier Science Publishers B.V., Amsterdam, The Netherlands.

TOLDRÁ, F., SANZ, Y. and FLORES, M. 2001. Meat Fermentation Technology. *In* Meat Science and Applications, Y.H. Hui, H. Shorthose, O. Young, M. Koohmaraie and R.W. Rogers (eds.), pp. 537–561. Marcel Dekker Inc., New York.

CHAPTER 9

NUTRITIONAL PROPERTIES

There is an interest for information on nutrient content of meat products. A great number of households in the U.S. and European Union include some meat products in their diets. These foods are usually eaten regularly and in significant quantities in certain countries where they are traditionally consumed. Thus, these foods may have some impact on total nutrient intake. Main compositions of dry-cured meat products are shown in Table 9.1. Dry-cured meat products have the same proportion as the starting meat but, because they have lost a substantial amount of water, these products have a much higher concentration in nutrients (nutrient density) (Schweigert 1987). Ash content depends on the formulation but is substantially increased due to salt addition. Breeding, feeding and environmental conditions have a strong influence on the composition of the meat products. Other ingredients and the method of preparation may also influence the composition of these products. For instance, sausages contain more fat and less protein than muscle.

PROTEINS

Animal proteins are considered to have high biological value (HBV) because they contain a high percentage of the essential amino acids required for growth and maintenance of body tissues and are present in very good proportions to meet the nutritional requirements. On the contrary, plant protein sources are limited in several essential amino acids such as lysine and sometimes threonine and tryptophan. Requirements for proteins vary depending on the physiological status and age. So, protein allowances for dietary proteins by FAO/WHO/UNU (Linder 1985) are as follows: 2.25–1.17 g/Kg/day for infants (0–2 years), 1.13–0.99 g/Kg/day for children (2–10 years), 1.00–0.86 g/Kg/day for adolescents (10–18 years) and 0.75 g/Kg/day for adults (over 18 years).

During the processing of dry-cured meat products, there is a continuous generation of free amino acids. These free essential amino acids are very important in poor nutritional quality diets or when the energy intake is low. This is also important for population sectors with specific needs, like children, elderly and diseased people. The generation of free amino acids in dry-cured ham is incredibly high (Table 9.2) as a result of an intense proteolysis (Toldrá *et al.* 1995, 2000). There are very large increases (several hundreds of mg/100 g) of lysine, alanine, leucine, aspartic and glutamic acids, valine, proline, threonine, serine, arginine, phenylalanine and tyrosine in all types of hams. The large

TABLE 9.1.
EXAMPLE OF PROXIMATE COMPOSITION OF SOME TYPICAL DRY-CURED MEAT PRODUCTS

Product type	Moisture (%)	Protein (%)	Fat (%)	Carbohydrate (%)	Ash (%)
American country-style, butt[1]	50.2	26.9	14	0.3	2.5
American country-style, shank[1]	57.4	23.6	10	1.1	2.2
Italian dry-cured ham, lean[2]	64.6	26.8	1.8	—	6.7
Spanish dry-cured ham, lean[3]	59.0	31.0	2.9	—	7.5
French dry-cured ham[4]	55.7	26.8	8.4	0.4	6.9
Italian dry-fermented sausage[5]	44.3	26.4	24.1	—	5.2
Spanish dry-fermented sausage[3]	32.9	22.3	36.3	0.7	5.6
French dry-fermented sausage[4*]	35.0	26.0	35.0	1.1	3.0
Pepperoni[6]	28.5	18.1	42.9	—	5.1

[1]Chen *et al.* 1997, [2]Bellatti *et al.* 1985, [3]Toldrá, unpublished data 1991, 1996, [4]CIC 1995, [5]Novelli *et al.* 1998, [6]Acton 1977
[*]in mg/100 g

TABLE 9.2.
EXAMPLE OF GENERATION OF FREE AMINO ACIDS (EXPRESSED IN MG/100 G) IN THE MUSCLE *BICEPS FEMORIS* OF DRY-CURED HAM AND IN DRY-CURED LOIN AND DRY-FERMENTED SAUSAGE

Amino acids	**Dry-cured ham**	**Dry-cured loin**	**Dry-fermented sausage**
Aspartic acid	319.4	16.7	43.4
Glutamic acid	620.4	231.0	180.6
Serine	264.3	56.1	49.4
Asparagine	60.4	14.0	22.5
Glycine	248.6	45.3	71.9
Glutamine	2.7	15.3	32.5
Tyrosine	167.0	36.4	4.4
Proline	254.1	42.2	47.8
Alanine	411.2	106.2	137.7
Arginine	375.6	28.2	64.4
Essential			
Threonine	232.4	44.2	50.1
Valine	328.4	76.6	94.0
Methionine	144.2	44.1	54.2
Isoleucine	253.6	66.5	75.1
Leucine	398.6	121.3	138.9
Phenylalanine	232.4	71.6	83.3
Tryptophan	53.8	15.1	13.3
Histidine	215.9	23.9	28.9
Lysine	702.5	79.4	58.1

(Adapted from Flores *et al.* 1998)

increase in free lysine (>700 mg/100 g) indicates a high degree of digestibility of ham proteins. The amount of taurine is also quite high (around 80 mg/100 g) and interesting because it plays an important role in nerve function (Gaull 1990). Glutamine is also relevant to different metabolic processes and may be important in preventing and/or treating diseases (Neu *et al.* 1996). In the case of dry-fermented sausages the increase in free amino acids is lower due to the shorter processing time and acid pH far from the optimal neutral pH for aminopeptidases.

Some nitrogen compounds also have an important function during the processing of meat and meat products. This is the case of several endogenous antioxidants such as carnosine and anserine, histidine-containing dipeptides that can contribute to the inactivation of lipid oxidation catalysts and/or free radicals in the cytosolic environment (Decker and Crum 1993). The result is a reduction in oxidative rancidity and stabilization of the color (Chan and Decker 1994). The concentrations of carnosine and anserine are higher in muscles with glycolytic activity like *M. Semimembranosus* and lower in oxidative muscles. These dipeptides are resistant to endogenous proteases and have shown a very good stability during the dry-curing process (Toldrá *et al.* 2000).

LIPIDS

Lipids include three groups of compounds: glycerides consisting in mono-, di-, and triglycerides esterified to one, two or three fatty acids, respectively; phospholipids; and cholesterol. Since meat products are generally considered high-fat foods, a substantial reduction in the fat content has been developed in recent years through modern breeding and feeding practices and through increased trimming of visible fat in the processing of the products (Godber 1994). The lean pork meat product, like dry-cured ham, does not contribute to an excess of calories since the amount of fat is below 30 g/Kg (Gandemer 1999). An excess of energy in the feed may result in an increased carcass fatness. On the other hand, diets rich in protein and lysine result in a lower fat deposition. Regarding the balance of fatty acids, which has received the most attention in recent decades due to its influence on the development of cardiovascular diseases, the amount of monounsaturated fatty acids is around 40–50% (Table 9.3) with oleic acid as the main fatty acid in both muscle and backfat tissues. This is very important since oleic acid is considered a very good fatty acid for raising the level of high-density lipoproteins (or HDL cholesterol) in plasm. The M/S ratio varies between 1.0–1.5. However, the problem of animal fats is that, whatever the calculation, they supply a large amount of saturated fatty acids. The P/S ratio ranges between 0.20 and 0.40 because linoleic and linolenic acids are those showing greater variations depending on the dietary treatment. It is well known that fatty acid composition, especially in depot fat, is strongly affected by dietary lipid composition (Rhee 1992; Cobos *et al.* 1994). Oleic and linoleic acids are incorporated more efficiently than saturated fatty acids such as palmitic or stearic acids (Mordenti *et al.* 1994).

TABLE 9.3.
EXAMPLE OF PROXIMATE LIPID COMPOSITION OF SOME TYPICAL DRY-CURED MEAT PRODUCTS

Product type	SFA (%)	MUFA (%)	PUFA (%)	Cholesterol (mg/100 g)
American country-style, butt[1]	37.1	62.9 (UFA)	—	110
American country-style, shank[1]	36.4	63.6 (UFA)	—	93
Italian dry-cured ham[2]	33.5	54.1	11.5	—
Spanish dry-cured ham[3]	35.5	51.0	11.8	65
French dry-cured ham[4]	38.3	50.3	11.4	66
Italian dry-fermented sausage[5]	37.6	43.5	12.5	70
Spanish dry-fermented sausage[3]	30.4	46.2	23.3	—
French dry-fermented sausage[4]	39.7	47.4	12.9	91

[1]Chen *et al.* 1997, [2]Baldini *et al.* 1992, [3]Navarro *et al.* 1997, [4]CIC 1995, [5]Novelli *et al.* 1998, Demeyer *et al.* 2000

The quality of the fat in the meat products will strongly depend on the feed used during fattening. For instance, the presence of soybean or corn oils raises the linoleic acid content in the fat, replacing the oleic acid to a large extent (higher C18:2/C18:1 ratio) (Monahan *et al.* 1991). Thus, these fats are more susceptible to iron-induced lipid peroxidation and then to the generation of oxidation products during the dry-curing process. The amount of polyunsaturated fatty acids is higher in those products rich in phospholipids because they are characterized by a high level of PUFA (Leseigneur-Meynier and Gandemer 1991). Linolenic acid is the major n-3 acid, followed by C20:5, C22:5 and C22:6, which are present at low percentages (less than 0.4%). Linoleic and arachidonic acids are the major n-6 acids.

Older pigs (cases of Iberian or Corsican pigs) yield higher intramuscular fat content and the resulting dry-cured meat products contain higher amounts of fats with a higher ratio in mono/polyunsaturated fatty acids than respective products manufactured from standard white pigs. The color is usually more intense which might be due to a higher accumulation of myoglobin or to a higher iron content of the ground where these pigs were reared outdoors in extensive systems (Gandemer 2001, personal communication).

The cholesterol content of the dry-cured meat products is, in general, not affected by dietary treatment (Busboom *et al.* 1991). Aportation of cholesterol is in the range of 50–70 mg/100 g for both lean or fat constituents of the meat product. The generation of cholesterol oxidation products is very low in dry-cured meat products. For instance, very small amounts of 5,6 α-epoxycholesterol and 7ß-hydroxycholesterol have been detected in some dry-fermented sausages at levels below 1 μg/g (Novelli *et al.* 1998; Demeyer *et al.* 2000).

VITAMINS

The dry-curing process itself has little effect on vitamins. Some loss in vitamins, especially in thiamine, may occur when the product is smoked. This loss can reach up to 50% if the product is at neutral pH and depending on the severity of heat processing (intensity and time of heating). Something similar happens with vitamin B_6, which is also sensible to heat. Some losses in riboflavin, niacin and vitamin B_{12} have also been observed depending on the intensity of heating, and small amounts of riboflavin and niacin may be initially lost in the drippings. Folates, usually present in reduced forms of pyroglutamates, are sensitive to oxidation. Finally, pantothenic acid is relatively stable (Reig and Toldrá 1998).

Water-Soluble Vitamins

In general, dry-cured meat products have a mild thermal treatment and thus constitute a good source of water-soluble vitamins. Furthermore, its concentrations appear to be higher than fresh pork meat because of moisture loss during drying. Water-soluble vitamins are described in detail (Reig and Toldrá 1998).

Thiamine. The active form present in the meat is thiamine pyrophosphate (TPP), which acts as a coenzyme in three enzymatic reactions catalyzed by pyruvate decarboxilase, α-ketoglutarate decarboxylase and transketolase. Beriberi is a well-known disease produced by a deficiency in thiamine. Requirements for thiamine are recommended around 0.50 mg/1000 kcal for adults, although they may be higher in the presence of antagonists, such as tea, coffee or with high intakes of carbohydrates. Dry-cured meat products constitute excellent sources of thiamine, around 0.9-1.0 mg/100 g (Calkins 1988; CIC 1995). These high values are mainly due to the presence of pork meat in the product since this meat is the richest source in thiamine compared to other meats or animal origin products.

Riboflavin or Vitamin B_2. This vitamin forms part of two important coenzymes: riboflavin 5′-phosphate or flavin-adenine dinucleotide (FAD), which are essential in all aspects of metabolism. Flavins are especially important in biological systems because they are stronger oxidizing agents than NAD^+, can participate in reactions involving one or two electrons and the reduced form can react directly with oxygen. The deficiency in riboflavin is in the frame of general poor diets and is associated with other vitamin B deficiencies. The RDA for healthy adults is 0.6 mg/1000 kcal, although these requirements do not increase with increased energy intake. Dry-cured meat products contain around 0.3 mg/100 g (CIC 1995) and constitute an important source of riboflavin.

Niacin or Vitamin B_3. This vitamin includes nicotinic acid and nicotinamide. Niacin is involved in many metabolic reactions such as the anaerobic glycolytic pathway, oxidative phosphorylation and fatty-acid biosynthesis and oxidation. Daily requirements are linked to the amount of tryptophan intake in the diet because niacin may be synthesized from dietary tryptophan (in a 1/60 rate). These are expressed as niacin equivalents. Weakness, loss of appetite and, in severe cases, pellagra are some of the symptoms associated with niacin deficiency. The needs for niacin also increase with calorie consumption. The RDA is 6.6 mg niacin equivalents/1000 kcal for adults. Dry-cured meat products are rich in tryptophan, around 0.75 mg/100 g, as a potential source of niacin and constitute a good source of niacin, around 7–8 mg/100 g (CIC 1995).

Pantothenic Acid or Vitamin B_5. This vitamin is present in free form or as coenzyme A. It forms part of two coenzymes: coenzyme A and phosphopantetheine that are used by more than 70 enzymes. It plays an important role in the reactions of energetic metabolism and as a specific transport system in the biosynthesis reactions of lipid compounds. The deficiency is unusual and would be in the frame of general malnutrition symptoms. The RDA for adults is 5–10 mg/day.

Pyridoxine or Vitamin B_6. This vitamin occurs in three forms, pyridoxine, pyridoxamine and pyridoxal. It is converted, through the pyridoxal kinase, into pyridoxal phosphate, the coenzyme form that participates in a great number of reactions associated with amino acid metabolism such as transaminations, decarboxylations, breakdown of side chains, etc. The number of enzymatic reactions exceeds one hundred. This is an important vitamin since its deficiency in the diet may often cause important changes, among others, in the central nervous system and abnormal electroencephalograms. The content of vitamin B_6 is higher in cereal grains, fruits and vegetables. However, its bioavailability is higher for animal foods because this vitamin is in the glycosylated form. The RDA depends on the protein in the diet and other situations like pregnancy, lactation, excess of alcohol intake, etc. In general, the RDA is 1.7–2.0 mg/day for adult males and 1.4–1.6 mg/day for females. The content of vitamin B_6 in dry-cured meat products is around 0.5–0.6 mg/100 g (CIC 1995).

Cobalamin or Vitamin B_{12}. This vitamin is only present in foods of animal origin. Vitamin B_{12} is essential for the normal function of all cells. Its deficiency causes anemia and physical and neurological problems. Vitamin B_{12} acts as a coenzyme in the synthesis of succinyl-CoA and is also necessary in the transfer of methyl groups from homocysteine to methionine. Metabolism and function of vitamin B_{12} are closely related to folic acid. The RDA for adults is in the range 2–3 μg/day. Dry-cured meat products contain around 0.5–1.0 μg/100 g (CIC 1995).

Folic Acid. A great part of this vitamin is linked to a polymeric structure of glutamic acid. Its deficiency, usually related to low dietary intakes, contributes to weakness, sleeplessness, etc. Daily requirements are related to body growth and its availability for absorption. The RDA may vary depending on age, but it is around 200 μg/day for adults, assuming a 50% efficiency in the absorption process. Chronic deficiency in vitamin B_{12} may be masked by an excess of this vitamin, which explains why the upper limit is restricted. Dry-cured meat products contain variable amounts depending on the product, between 1.5 and 13 μg/100 g meat (CIC 1995).

Vitamin C. Ascorbic acid is a powerful reducing agent and functions as an antioxidant at the cellular level. The RDA is 60 mg/day, but these requirements are increased due to numerous factors such as tobacco, contraceptives, etc. Pork meat contains only negligible amounts of vitamin C, but the addition of ascorbates to the curing salt, reaching up to 500 mg/Kg in the product, enhances the content of vitamin C in dry-cured meat products.

Fat-Soluble Vitamins

The content in fat-soluble vitamins is reduced and, in some cases, almost negligible. These vitamins are described below (Reig and Toldrá 1998).

Vitamin A. Retinoids are formed in foods of animal origin. Night blindness, followed by conjunctival xerosis of the eyes, diarrhea and respiratory diseases appear when there are deficiencies in this vitamin. Dietary vitamin A activity is expressed as Retinol Equivalents (1 RE activity of 1 μg retinol). The RDA for an adult is around 1000 RE. Dry-cured meat products contain poor amounts of vitamin A, below 5 RE/100 g. However, its lipids and proteins contribute to a better absorption of plant carotenoids (Smith 1990).

Vitamin D. This vitamin is largely dependent on the season and/or time of exposure to the sunlight because it is synthesized in the skin when irradiated with UV light. Its deficiency reduces the intestinal absorption of calcium and phosphorus and often produce major deformation of the skeleton. The RDA for adults is around 200–400 IU (5–10 μg of vitamin D_3). Dry-cured meat products contain very small amounts of vitamin D, below 0.3 μg/100 g.

Vitamin K. The vitamin K_2 or menaquinone is produced by microorganisms in the gastrointestinal tract. The main function of vitamin K is in blood clotting. Deficiencies of vitamin K are very rare in humans. Around 1 μg/Kg body weight/day is recommended. Dry-cured meat products contain scarce amounts of vitamin K, below 10 μg/100 g (CIC 1995).

Vitamin E. This vitamin defines the biological activity associated with α-tocopherol that constitutes one of the most important endogenous antioxidants in muscle. It protects lipids, especially phospholipids, against the damaging effects of oxidation. Muscle is susceptible to oxidative deterioration because it contains lipid oxidation catalysts and membranes rich in phospholipids, which usually contain high proportions of unsaturated fatty acids. Vitamin E, a major lipid-soluble antioxidant in skeletal muscle, contains a phenolic structure that scavenges lipid and oxygen radicals (Chan and Decker 1994). This preserves both meat flavor and color (Buckley *et al.* 1995). Daily recommended requirements for adults are about 0.4 mg/g of PUFA. The amount of vitamin E in dry-cured meat products depends on animal diet because it is not synthesized by animal tissues. Supplementation of animal diets with α-tocopherol is being performed in many farms worldwide for minimizing the formation of secondary lipid oxidation products. In this sense, the concentration in the product may range from 3.2 μg α-tocopherol/100 g when fed a basal diet to 7.0–8.0 μg α-tocopherol/100 g when fed 200 mg α-tocopherol/Kg for 2 weeks (Chan and Decker 1994) or up to 30 μg α-tocopherol/100 g when fed for 3 months (Hoving-Bolink *et al.* 1998). The supplementation of α-tocopherol lowers lipid oxidation and color fading during the storage of sliced dry-cured ham and minimizes weight loss during storage (Isabel *et al.* 1999).

MINERALS

Two major groups of minerals can be distinguished, macrominerals and trace elements. Dry-cured meat products constitute a good source of both, especially taking into account that muscle foods, in general, promote its bioavailability (Godber 1994). The mineral components are mainly associated with the water and protein fractions. The main mineral compounds in meat products are described in detail.

Iron

Meat products are a rich source of iron. Furthermore, their contribution is even higher due to the increased bioavailability (Godber 1994). Red meats contain a high proportion of heme iron, between 50–60% usually from myoglobin and some remaining hemoglobin, that is efficiently absorbed in our organism. Several factors enhance nonheme iron absorption. Meat proteins significantly increase this absorption. Ascorbic acid, used as a curing adjunct in dry-cured meat products, also exerts an enhancing effect when consumed in the same meal and at a rate of 100 mg ascorbic acid:3 mg iron (Worthington *et al.* 1990). So, dry-cured meat products may also promote the bioavailability of nonheme iron from plant foods that are eaten with the meat product. Deficiencies in iron result in anemia and can impair intellectual performance, behavior, resistance to infections and work performance (Johnson and Fischer 1994). Some groups at risk are preschool children, women in their reproductive years and adolescents. Numerous nutritional education programs have been addressed to these populations.

The RDA depends on several variables but depends especially on individual differences in absorptive capacity and different iron bioavailability among foods. So, the RDA for adult males is 9, 6 and 5 mg/day when diets contain less than

10%, 10–25% or higher than 25%, respectively, of calories from animal foods. These recommendations increase up to 28, 19 and 14 mg/day, at the respective same conditions for menstruating women (Worthington *et al.* 1990).

Zinc

This trace element is essential for its implication in many essential cell functions. Several zinc metalloenzymes need zinc for catalytic, structural, regulatory or modulatory activity. Other zinc-dependent enzymes mediate DNA replication, transcription and translation. Zinc is also essential for development and function of the immune system and brain, structure and function of cell membranes, appetite control, gustatory function and skeletal development. Deficiencies in zinc may result in the Prasad-Halsted syndrome of growth stunting and hypogonadism. Other symptoms may appear in individuals with conditioned zinc deficiency due to poor absorption, use or retention of zinc because of medication, illness or behavior (Sandstead *et al.* 1990). The RDA is around 10–20 mg/day but depends on the bioavailability, the level of anabolism and losses by excretion. Meat products contain around 3–5 mg/100 g (Calkins 1988). The presence of meat enhances the bioavailability of zinc, which is a major provider of dietary zinc.

Selenium

This natural trace element has essential functions in living organisms, especially in protection against free-radical injury. It is a constituent of glutathione peroxidase, an enzyme located in mitochondria and cytosol of skeletal muscle cells with antioxidant activity in our organism (Chan and Decker 1994). Selenium is necessary for the activity of this enzyme. Deficiency in selenium is associated with chronic diseases such as those observed in certain regions of China. This deficiency may cause cardiomyopathy, an increased risk of cardiovascular diseases and primary liver cancer (Johnson and Fischer 1994). The recommended amounts are 0.055 mg/day for women and 0.070 mg/day for men, not far from the amount of selenium in pork meat which is around 0.015–0.030 mg/100 g (Venäläinen *et al.* 1997).

Copper

Copper is a cofactor of many oxidative enzymes and is essential for muscle metabolism, respiratory metabolism, destruction of free radicals, etc. Copper is required for heart function, bone formation, energy metabolism, nerve transmission, elastin synthesis, skin pigmentation, etc. (Johnson and Fischer 1994). The requirement for adults is 1.5–3.0 mg/day. Deficiencies in copper may result in bone fractures, hypercholesterolemia, low glucose tolerance, blood pressure aberrations, etc. (Johnson and Nielsen 1990). Meat products contain 0.08–0.09 mg/100 g (Calkins 1988).

Manganese

This trace element participates in numerous enzymatic reactions as catalyst, cofactor or constituent of enzymes involved in protein and energy metabolism, regulation of carbohydrate and lipid metabolism and cartilage and bone formation (Johnson and Nielsen 1990). Deficiencies may produce decreased serum cholesterol and a fleeting dermatitis. The Estimated Safe and Adequate Intake by the Food and Nutrition Board (1980) is 2.5–5 mg/day. The content of manganese in dry-cured meat products is 0.01–0.05 mg/100 g (CIC 1995).

Sodium

The sodium content in dry-cured meat products is high due to the addition of curing salts during processing. The amount of salt may be as high as 8 g/100 g in certain dry-cured hams, being sensibly lower in dry-fermented sausages (< 3 g/100 g). About 30% of normal individuals and 50% of hypertensive individuals are salt-sensitive (Karanja *et al.* 1990). Consumption of dry-cured meat products is thus not recommended for individuals with dietary salt restriction.

REFERENCES

ACTON, A.C. 1977. The chemistry of dry sausages. Proc. Annual Reciprocal Meat Conf. *30*, 49–62.

ARISTOY, M-C. and TOLDRÁ, F. 1993. Availability of essential amino acids in dry-cured ham. Int. J. Food Sci. Nutr. *44*, 215–219.

ARISTOY, M-C. and TOLDRÁ, F. 1998. Concentration of free amino acids and dipeptides in porcine skeletal muscles with different oxidative patterns. Meat Sci. *50*, 327–332.

BALDINI, P., BELLATTI, M., CAMORALI, G., PALMIA, F., PAROLARI, G., REVERBERI, M., PEZZANI, G., GUERRIERI, C., RACZYNSKI, R. and RIVALDI, P. 1992. Characterization of Italian raw ham by chemical, physical, microbiological and organoleptic parameters. Ind. Conserve *67*, 149–159.

BELLATI, M., DAZZI, G., CHIZZOLINI, R., PALMIA, F. and PAROLARI, G. 1985. Viandes Prod. Carn. *6*, 142–145.

BUCKLEY, D.J., MORRISSEY, P.A. and GRAY, J.I. 1995. Influence of dietary vitamin E on the oxidative stability and quality of pig meat. J. Anim. Sci. *73*, 3122–3130.

BUSBOOM, J.R., RULE, D.C., COLIN, D., HEALD, T. and MAZHAR, A. 1991. Growth carcass characteristics and lipid composition of adipose tissue and muscle of pigs fed canola. J. Anim. Sci. *69*, 1101–1108.

CALKINS, C.R. 1988. Pork in the human diet. *In* New Ideas. A Pork Technical Reference Manual. National Livestock and Meat Board. Vol. 2, 5, 3–24.

CHAN, K.M. and DECKER, E.A. 1994. Endogenous skeletal muscle antioxidants. CRC Crit. Rev. Food Sci. Nutr. *34*, 403–426.

CHEN, H-Y., PILKINGTON, D.H., THARRINGTON, J.B. and ALLEN, J.C. 1997. Developing a dry-cured ham nutritional database. J. Food Comp. Anal. *10*, 190–204.

CIC. 1995. Nutrition, Santé et charcuteries. Centre d´information sur les charcuteries. Paris, France, 1–48.

COBOS, A., DE LA HOZ, L., CAMBERO, M.I. and ORDOÑEZ, J.A. 1994. Review: Influence of animal diet on fatty acids in meat lipids. Rev. Esp. Ciencia Tecnol. Aliment. *34*, 35–51.

DECKER, E.A. and CRUM, A.D. 1993. Antioxidant activity of carnosine in cooked ground pork. Meat Sci. *34*, 245–253.

DEMEYER, D. *et al.* 2000. Control of bioflavour and safety in fermented sausages: first results of a European project. Food Research Int. *33*, 171–180.

FLORES, M., SANZ, Y., SPANIER, A.M., ARISTOY, M.C. and TOLDRÁ, F. 1998. Contribution of muscle and microbial aminopeptidases to flavor development in dry-cured meat products. *In* Food Flavor: Generation, Analysis and Process Influence, Contis *et al.* (eds.), pp. 547–557. Elsevier Science B.V., Amsterdam, The Netherlands.

GANDEMER, G. 1999. Lipids and meat quality: Lipolysis, oxidation, flavor. Sci. Aliments *19*, 439–458.

GODBER, J.S. 1994. Nutritional value of muscle foods. *In* Muscle Foods, D.M. Kinsman, A.W. Kotula and B.C. Breidenstein (eds.), pp. 430–455. Chapman and Hall, New York.

GAULL, G.E. 1990. Taurine in pediatric nutrition: a review and update. Pediatrics *83*, 433–442.

HOVING-BOLINK, A.H., EIKELENBOOM, G., DIEPEN, T.M., JONGBLOED, A.W. and HOUBEN, J.H. 1998. Effect of dietary vitamin E supplementation on pork quality. Meat Sci. *49*, 205–212.

ISABEL, B., LÓPEZ BOTE, C.J., REY, A.I. and SANZ, R. 1999. Influence of dietary α-tocopheryl acetate supplementation of pigs on oxidative deterioration and weight loss in sliced dry-cured ham. Meat Sci. *51*, 227–232.

JOHNSON, M.A. and FISCHER, J.G. 1994. Role of minerals in protection against free radicals. Food Technol. *48*, 112–120.

JOHNSON, P.E. and NIELSEN, F.H. 1990. Copper, manganese, cobalt and magnesium. *In* Meat and Health, A.M. Pearson and T.R. Dutson (eds.), pp. 275–299. Elsevier Applied Science, London, UK.

LESEIGNEUR-MEYNIER, A. and GANDEMER, G. 1991. Lipid composition of pork muscle in relation to the metabolic type of the fibres. Meat Sci. *29*, 229–241.

LINDER, M-C. (ed.) 1985. Nutritional biochemistry and metabolism with clinical applications. Elsevier Science, Amsterdam, The Netherlands, 81.

McCAIN, G.R., BLUMER, T.N., GRAIG, H.B. and STEEL, R.G. 1968. Free amino acids in ham muscle during successive aging periods and their relation to flavor. J. Food Sci. *33*, 142–145.

MONAHAN, F.J., BUCKLEY, D.J., MORRISEY, P.A., LYNCH, P.B. and GRAY, J.I. 1991. Influence of dietary fat and α-tocopherol supplementation on lipid oxidation in pork. Meat Sci. *29*, 229–241.

MORDENTI, A., PIVA, G. and DELLACASA, G. 1994. Nutrition and fat quality in the heavy pig. Ital. J. Food Sci. *2*, 141–156.

NAVARRO, J.L., NADAL, M.I., IZQUIERDO, L. and FLORES, J. 1997. Lipolysis in dry-cured sausages as affected by processing conditions. Meat Sci. *45*, 161–168.

NEU, J., SHENOY, V. and CHAKRABARTI, R. 1996. Glutamine nutrition and metabolism: where do we go from here? FASEB J. *10*, 829–837.

NOVELLI, E., ZANARDI, E., GHIRETTI, G.P., CAMPANINI, G., DAZZI, G., MADARENA, G. and CHIZZOLINI, R. 1998. Lipid and cholesterol oxidation in frozen stored pork, salame Milano and mortadella. Meat Sci. *48*, 29–40.

REIG, M. and TOLDRÁ, F. 1998. Pork meat as a source of vitamins. Recent Res. Devel. Nutr. Res. *2*, 29–34.

RHEE, K.S. 1992. Fatty acids in meat and meat products. *In* Fatty Acids in Foods and Their Health Implications, C.K. Chow (ed.), pp. 65. Marcel Dekker, New York.

SANSTEAD, H.H., DARNELL, L.S. and WALLWORK, J.C. 1990. Role of zinc and the contribution of meat to human nutrition. *In* Meat and Health, A.M. Pearson and T.R. Dutson (eds.), pp. 237–274. Elsevier Applied Science, London, U.K.

SMITH, J.E. 1990. Fat-soluble vitamins: Vitamins A, D, E and K. *In* Meat and Health, A.M. Pearson and T.R. Dutson (eds.), pp. 371–400. Elsevier Applied Science, London, U.K.

SCHWEIGERT, B.S. 1987. The nutritional content and value of meat and meat products. *In* The Science of Meat and Meat Products, J.F. Price and B.S. Schweigert (eds.), pp. 275–305. Food & Nutrition Press Inc., Trumbull, CT.

TOLDRÁ, F., FLORES, M. and ARISTOY, M-C. 1995. Enzyme generation of free amino acids and its nutritional significance in processed pork meats. *In* Food Flavors: Generation, Analysis and Process Influence, G. Charalambous (ed.), pp. 1323–1344. Elsevier Science, Amsterdam, The Netherlands.

TOLDRÁ, F., REIG, M., HERNÁNDEZ, P. and NAVARRO, J.L. 1996. Lipids from pork meat as related to a healthy diet. Recent Res. Devel. Nutr. *1*, 79–86.

TOLDRÁ, F. and REIG, M. 1998. Protein nutritional quality of muscle foods. Recent Res. Devel. Agric. Food Chem. *2*, 71–78.

TOLDRÁ, F., ARISTOY, M-C. and FLORES, M. 2000. Contribution of muscle aminopeptidases to flavor development in dry-cured ham. Food Res. Int. *33*, 181–185.

VENÄLÄINEN, E-R., HIRVI, T. and RIN, J. 1997. Effect of selenium supplementation on the selenium content in muscle and liver of Finnish pigs and cattle. J. Agric. Food Chem. *45*, 810–813.

WORTHINGTON-ROBERTS, B. and MONSEN, E.R. 1990. Iron. *In* Meat and Health, A.M. Pearson and T.R. Dutson (eds.), pp. 185–235. Elsevier Applied Science, London, U.K.

CHAPTER 10

EFFECT OF RAW MATERIALS AND PROCESSING ON QUALITY

Hams intended for processing are highly heterogeneous. They are obtained from pigs differing in genetic type, slaughtering age, morphology, lean and fat distribution, etc. The situation is even worse in the case of meat as raw material for dry-fermented sausages. In addition, there are many factors with strong influences on the process, which therefore affect the final quality of the products. This prompted several studies to evaluate the suitability of specific genetic lines and corresponding crosses for the production of dry-cured meat products with optimal quality. Today, it is known that raw meat (genotype, age, sex, ante and postmortem treatment) and process technology have a decisive influence on the activity of muscle enzymes as will be extensively discussed in this chapter. Furthermore, the presence of exogenous enzymes from microbial origin, which is the case in dry-fermented sausages, also contribute to modulate the final quality of the product. Processing technology also exerts an important influence on the enzyme activity and stability.

IMPORTANCE OF RAW MATERIALS

Effect of the Genetic Type

Current pig breeding schemes are usually based on a backcross or on a three- or four-way cross. For instance, the most common cross in Spain is a three-way cross where the sow is an F1 Landrace × Large White (LR × LW) crossbreed. The choice of the terminal sire depends on the profitability obtained per animal and, in this sense, slaughterhouses play an important role in this selection. The pig carcass price is fixed according to an evaluation consisting of a good score when the backfat is reduced and the conformation provides a high percentage of valuable cuts. So, sows are crossed either with a heavily muscled sire, like Pietrain (Pi) or Belgian Landrace (BL), or with a good growth rate and resistance breed like Duroc (DU).

Several studies show the effect of genetic type on meat quality (Oliver *et al.* 1994; Guerrero *et al.* 1996; Armero *et al.* 1999c). From these studies, it can be concluded that the Duroc breed provides high meat quality with a good intramuscular fat level appreciated for dry-cured meat products. Additionally, the cross with Duroc terminal sire grows faster and shows a better food conversion ratio (Blasco *et al.* 1994). Belgian Landrace and Pietrain have a high susceptibility to stress and thus have a high incidence of exudative meats. In these cases, a higher intensity in pastiness, crumbliness, brightness and nonprotein nitrogen and tyrosine content in dry-cured hams have been reported (Guerrero *et al.* 1996). Belgium Landrace crossbred pigs tend to have a higher percentage of discarded hams during processing and tend to have lower ham flavor scores (Gallo *et al.* 1994). An intermediate situation, combining good conformation and meat quality, can be obtained with the Belgian Landrace × Landrace (BL × LR) cross.

Crossbreeds have a definite effect on the intramuscular fat content but are restricted on the content of volatile compounds and flavor (Berdagué *et al.* 1993). Only fat aroma is influenced by the pig crossbreed, but this is due to the fat content. In studies of Italian crossbreeds with different percentages of Duroc and Large White, there were close relations between ham weight and cured ham moisture, marbling and muscle firmness in the finished product and between proteolytic enzyme activity in raw and cured hams (Schivazzappa *et al.* 1998). Even though some breeds modify the chemical composition of processed meat, the observed changes are sometimes enough for a substantial effect on the quality and organoleptic properties of dry-cured meat products.

Crossbreeding of traditional breeds (Large White, Landrace and Duroc) with different percentages of breeds such as Duroc or Landrace to produce pigs for ham industrial processing have, in general, few effects on curing suitability, processing quantitative traits and chemical and quality properties of dry-cured ham. Significant differences for all primary and secondary cuts are usually found for different sired pigs. For instance, the Belgian Landrace sired pigs have the highest proportion of valuable cuts such as ham, shoulder and chops while Duroc sired pigs tend to have higher proportion of ham and shoulder and lower of ribs (Blasco *et al.* 1994; Armero *et al.* 1999c). On the other hand, carcass weight also affects the yield of several main cuts. Therefore, shoulder, ribs and chops have been observed to increase and bacon trimmings and backfat to decrease with carcass weight (Armero *et al.* 1999c).

The enzymatic fingerprints and quality traits of five different crossbreeds, generally used for dry-cured ham, are shown in Fig. 10.1 and Fig. 10.2, respectively. The meat quality traits are important to evaluate the incidence of exudative meats for each crossbreed as well as color and water-holding capacity. The enzyme fingerprints that include endo- and exoproteases as well as lipases and esterases are useful for evaluating the expected proteolysis and lipolysis during the process and the potential benefits to meat quality such as tenderness and flavor development (Armero *et al.* 1999a, b). For instance, Belgian Landrace breed suggests a low aptitude of its meat to provide dry-cured hams based on the role of the exopeptidases and free generating flavor precursors (Armero *et al.* 1999b). This crossbreed, which has a high susceptibility to stress, also showed the largest PSE incidence at around 50% of the pigs. On the

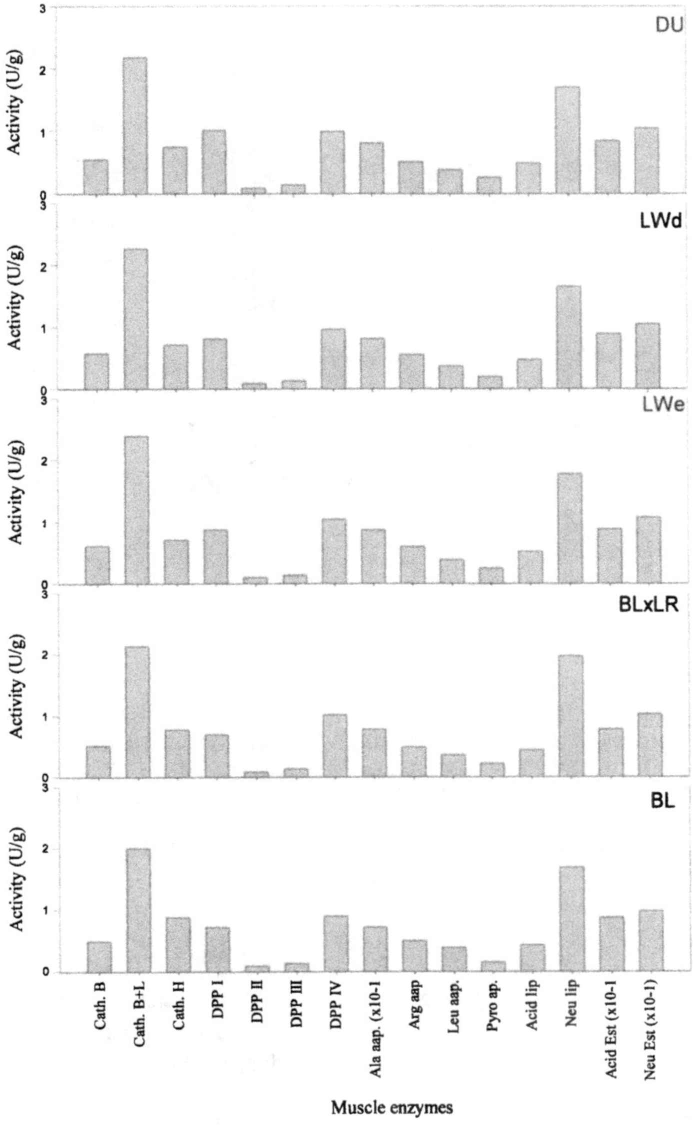

FIG. 10.1. ENZYME PROFILE IN THE MUSCLE *BICEPS FEMORIS* FROM THE OFFSPRING OF LANDRACE × LARGE WHITE CROSSBREED SOWS MATED WITH DANISH DUROC (DU), DUTCH LARGE WHITE (LWD), ENGLISH LARGE WHITE (LWE), BELGIAN LANDRACE × LANDRACE (BL × LR), BELGIAN LANDRACE (BL). ONE UNIT OF ENZYME ACTIVITY IS DEFINED AS THE AMOUNT OF ENZYME CAPABLE TO HYDROLYZE 1 μMOL OF SUBSTRATE PER HOUR AT 37C

(Adapted from Armero *et al.* 1999a, b)

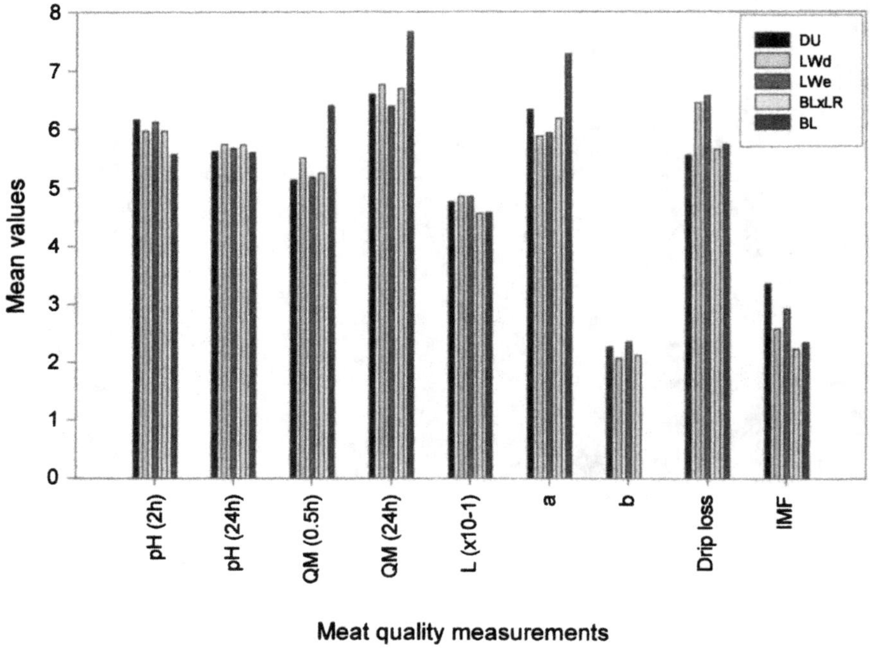

FIG. 10.2. COMPARISON OF QUALITY TRAITS FROM THE OFFSPRING OF LANDRACE × LARGE WHITE CROSSBREED SOWS MATED WITH DANISH DUROC (DU), DUTCH LARGE WHITE (LWD), ENGLISH LARGE WHITE (LWE), BELGIAN LANDRACE × LANDRACE (BL×LR), BELGIAN LANDRACE (BL). THE TRAITS ARE: pH AT 2 AND 24 H, CONDUCTIVITY (QM) AT 0 AND 24 H, COLOR PARAMETERS (L, a AND b), DRIP LOSS AND INTRAMUSCULAR FAT CONTENT (IMF IN %)
(Adapted from Armero *et al.* 1999c)

other hand, the Duroc-sired pigs give good scores on carcass conformation and meat quality as well as on the presence of intramuscular fat (marbling), which is highly appreciated by the dry-cured ham industry (Armero *et al.* 1999c).

The comparison of Iberian pig breed with White crossbreed pig, consisting in (Large White × Landrace) × Duroc reveals large differences between muscle proteolytic and lipolytic enzymes (Fig. 10.3). In fact, Iberian pigs show higher amounts of cathepsin D, dipeptidylpeptidase III and alanyl aminopeptidase, while the White crossbreed pigs show higher amounts of calpain and cathepsins B, B+L and H; dipeptidylpeptidases II and IV; aminopeptidase B; leucyl and pyroglutamylaminopeptidase; and acid lipase and neutral esterase (Rosell and Toldrá 1998).

Effect of Age

The screen of the proteolytic and lipolytic enzyme activity in raw hams from light (7-8 months old) and heavy pigs (11 months old) gives important information (Fig. 10.4). The use of PCA reduces the large set of variables and reveals the different enzymatic patterns for light and heavy hams (Toldrá *et al.* 1996). Heavy hams are characterized by a greater peptidase to proteinase ratio and a higher lipase, DPP IV and pyroglutamyl aminopeptidase activity. On the other hand, light hams show two groups. The large one is higher in moisture content and cathepsins B and B+L and lower in peptidase activity, while the minor one is intermediate in cathepsin B activity and larger in peptidase activity. In general, a higher moisture content is usually related to a high level of cathepsin B and B+L (Schivazzappa *et al.* 1992). So, light hams with high cathepsin B activity and a low amount of added salt are more prone to

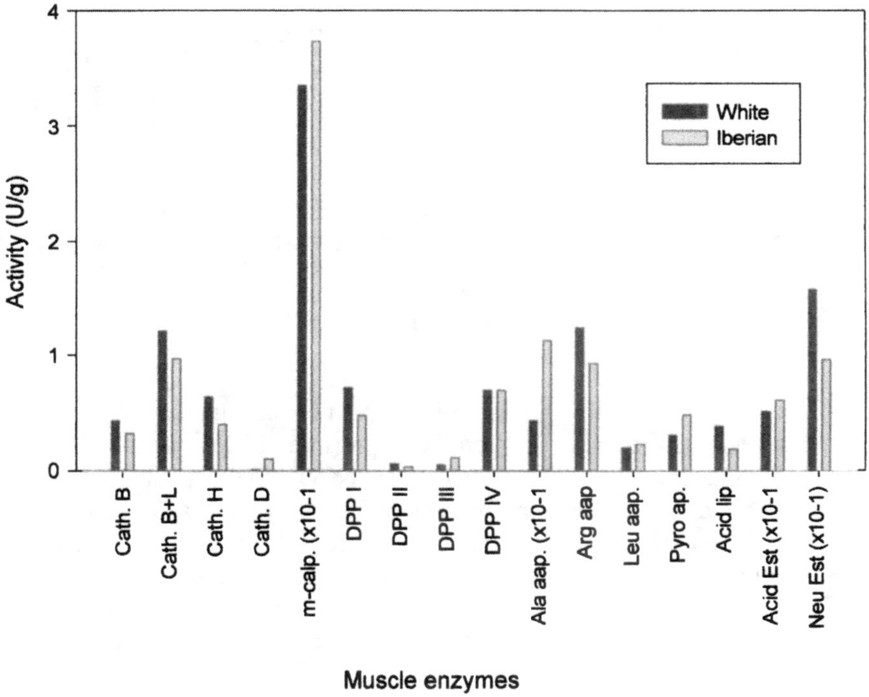

FIG. 10.3. COMPARISON OF ENZYME PROFILE IN THE MUSCLE *BICEPS FEMORIS* FROM WHITE AND IBERIAN PIGS. ONE UNIT OF ENZYME ACTIVITY IS DEFINED AS THE AMOUNT OF ENZYME CAPABLE TO HYDROLYZE 1 μMOL OF SUBSTRATE PER HOUR AT 37C

(Adapted from Rosell and Toldrá 1998)

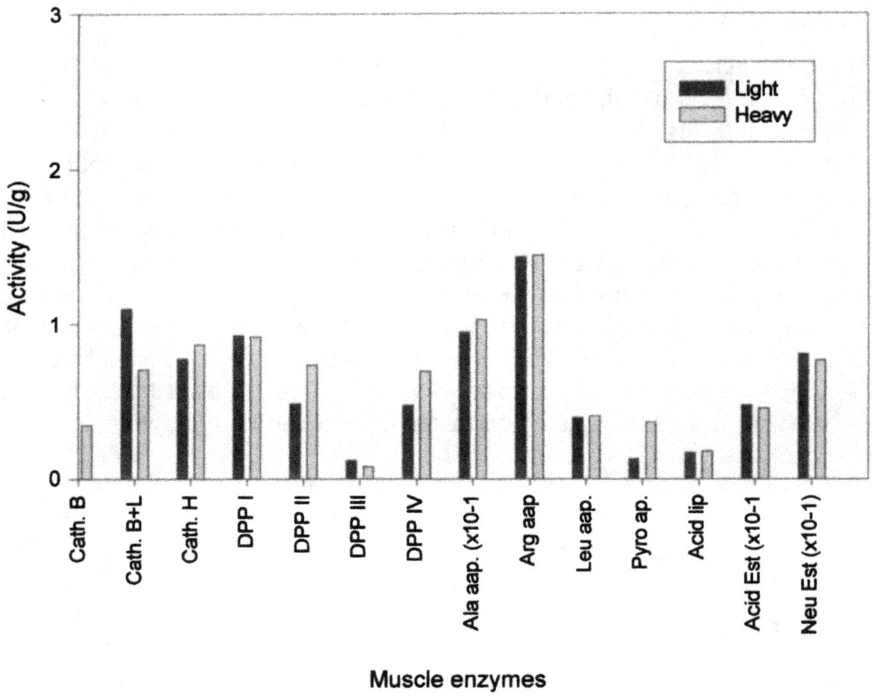

FIG. 10.4. COMPARISON OF ENZYME PROFILE IN THE MUSCLE *BICEPS FEMORIS* FROM LIGHT AND HEAVY PIGS. ONE UNIT OF ENZYME ACTIVITY IS DEFINED AS THE AMOUNT OF ENZYME CAPABLE TO HYDROLYZE 1 μMOL OF SUBSTRATE PER HOUR AT 37C
(Adapted from Toldrá *et al.* 1996)

proteolysis and thus give problems such as poor texture and mouthfeel (softness) and development of white film on the cut surface (Sárraga *et al.* 1993).

A higher intramuscular fat content may be observed with the increase in the age of the animal. In addition, the meat tends to be more flavorful and colorful due to an increased concentration of volatiles and myoglobin, respectively. Regarding the distribution of the main cuts in the carcass, the proportions of shoulder, ribs and chops decrease and the proportions of bacon trimmings and backfat increase when the carcass weight increases (Armero *et al.* 1999c).

Effect of Sex

Hams from barrows are fatter than those from gilts. They have a higher marbling and the subcutaneous fat layer is thicker, making oxygen diffusion more difficult. The higher marbling results in a slower salt diffusion and lower processing losses (Gou *et al.* 1995). The thicker layer of subcutaneous fat in barrows explains the higher processing loss in the gilts, even though both initially have a similar moisture content. In spite of the lower backfat thickness of gilt carcasses, there is no difference in subjective ham conformation scores (Blasco *et al.* 1994). Some differences in distribution of carcass weight may be found, but in general, gilts yield more ham, shoulder and loin. In the case of muscle enzymes, only minor differences have been found (Fig. 10.5).

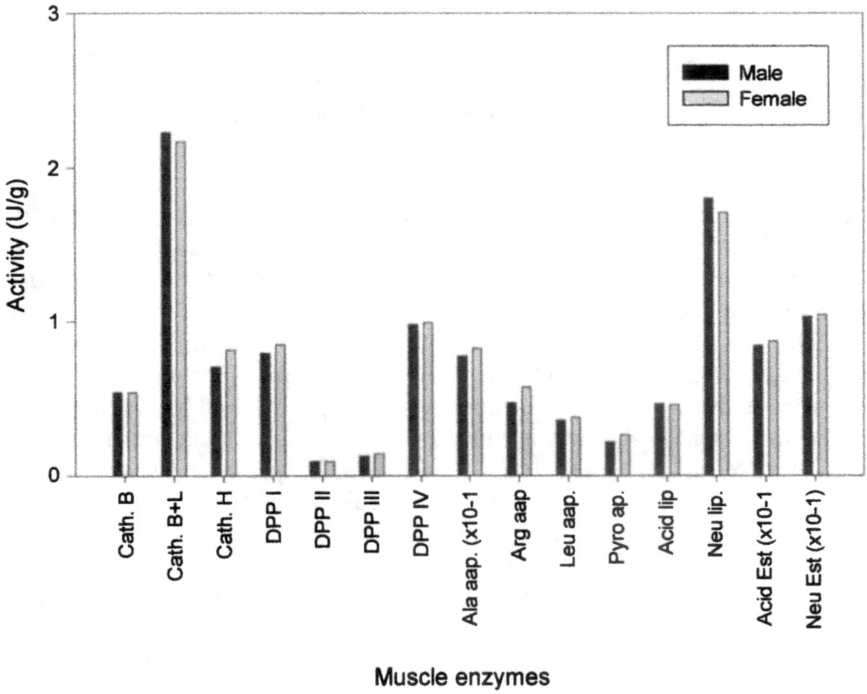

FIG. 10.5. EFFECT OF SEX ON THE MUSCLE ENZYME PROFILE. ONE UNIT OF ENZYME ACTIVITY IS DEFINED AS THE AMOUNT OF ENZYME CAPABLE TO HYDROLYZE 1 μMOL OF SUBSTRATE PER HOUR AT 37C
(Adapted from Armero *et al.* 1999a, b)

There is no sex effect on color measurements. Only a redder color is observed by panelists in gilts probably because of the higher marbling in the barrows (Gou *et al.* 1995). There is no noticeable difference in organoleptic properties of dry-cured ham from barrows and gilts (Gallo *et al.* 1994). A slightly high overall quality is found for hams from females because they have a higher fat complex and a lower salty taste (Oliver *et al.* 1994; Armero *et al.* 1999c). Hams from carcasses of barrows are preferred to avoid sexual odor problems due to high contents of androstenone or escatol detected in hams from entire males. Consumers wish for the best ham quality and are sensitive to boar taint sometimes found in dry-cured hams from noncastrated males (Diestre *et al.* 1990).

Effect of Feed Type

The composition of fatty acids, especially in depot fat, is strongly affected by feed. Pigs incorporate part of the dietary fatty acids into the fat tissue (Rhee 1992; Toldrá *et al.* 1996). The extent of incorporation may vary depending on the specific fatty acid and the type of feed. For instance, oleic and linoleic acids are incorporated more efficiently than palmitic or stearic acids. Feeds rich in saturated fats, such as tallow, give the highest levels of palmitic, palmitoleic, stearic and oleic acids in pork loin (Morgan *et al.* 1992). Other feeds rich in corn may increase the amount of linoleic acid. Different fatty acid profiles in the inter- and intramuscular fat of Iberian pigs have been observed when varying the proportions of cereals and acorn in the feed (Flores *et al.* 1988; De la Hoz *et al.* 1996). An example of the effect of the type of feed on the fatty acid composition of subcutaneous adipose tissue of Iberian pig is shown in Fig. 10.6. The acorn is mostly appreciated, although it is expensive and its production is dependent on many factors like climate, rain, etc. This is the reason for the addition of cereals and feeds, but this decreases the quality of the fat. In this sense, and due to the importance of fat for the final quality, the analysis of the fatty acid profile of subcutaneous and muscle fat may be used for predicting the feeding background and thus confirm the quality of the meat product obtained (Ruiz *et al.* 1998). Linoleic and linolenic acids usually show great variations, which may reach up to 40%, between the leanest and fattest animals (Enser *et al.* 1988). So, the addition of a high unsaturated oil, such as safflower oil, to the pig feed results in less palmitic and oleic acids and more 18:2, 20:2 and 20:3 acids (Larick *et al.* 1992). The excessive addition of oil in the feed, such as safflower, sunflower or canola, gives softer and oily fats that may result in unexpected oxidations during the dry-curing process and development of unpleasant flavor. Something similar may happen with an excess of soybean oil in the feed, which may produce an increase in the linoleic acid content and a replacement of the oleic acid to a large extent (Monahan *et al.* 1992).

The presence of fatty acids with high unsaturation may result in more intense oxidative processes and the possibility of rancidity development. To prevent these unpleasant oxidations, different studies have shown the protective effect of vitamin E. This compound is added in the feed as an antioxidant and is accumulated by pigs in tissues and subcellular structures, including membranes, increasing its effectivity substantially. Vitamin E is degraded in a minor way during the dry-curing process. However, the presence of salt as prooxidant, oxidative changes and long processing time help to increase the oxidative stability of the hams during the process (Isabel *et al.* 1998).

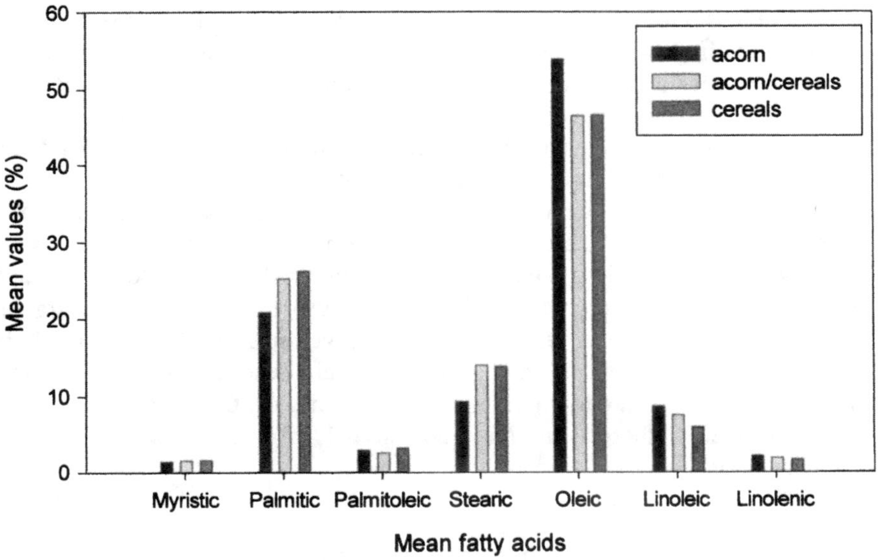

FIG. 10.6. EFFECT OF THE TYPE OF FEED (ACORN, MIXTURE OF ACORN AND CEREALS AND CEREALS) ON THE MEAN FATTY ACIDS OF SUBCUTANEOUS ADIPOSE TISSUE OF IBERIAN PIGS
(Adapted from Flores *et al.* 1998)

In general, the amount of intramuscular fat may be increased with a good level of nutrition. On the other hand, an induced lipolysis may be achieved with food deprivation, resulting in a higher content of free fatty acids and monoglycerides, especially in glycolytic muscles (Fernandez *et al.* 1995). This lipolysis can be significantly detected in white adipose tissue after 72 hours of food deprivation.

Effect of Antemortem Stress

Postmortem changes have considerable variation in rate and extent and are considerably influenced by ante and postmortem conditions. The rate and extent of pH fall in postmortem muscle influences the appearance and quality of the meat, which therefore affects the final quality of the dry-cured meat products. A rapid and deep decline in pH results in pale, soft and exudative (PSE) meat, which affects the quality perceived by the consumers and also by meat processors because of high losses in drippings during processing. A similar effect may be found with regular, soft and exudative (RSE) meat, which has a high drip loss but keeps a normal coloration. When the pH remains high, the meat results in dark, firm and dry (DFD) meat. This meat has a high water-holding capacity and is prone to microbial spoilage due to its neutral pH.

It is necessary to measure the pH at 1 hour postmortem for detection of exudative meats and at 24 hours for detection of DFD meats. Both exudative and DFD hams represent important processing problems in the dry-cured meat industry as well as in the final quality of the products. Exudative (PSE and RSE) hams have higher amounts of moisture on the surface due to the lower water-binding capacity, which helps the dissolution of the added solid salt and facilitates its penetration into the ham. They also have protein denaturation phenomena, a pale color and tend to have dryer surfaces that can result in muscles disjunctions where microorganisms could penetrate into the ham and spoil it. DFD hams have a higher moisture content, and the migration of salt is faster. However, these hams must be rejected in order to avoid microbial contamination.

INFLUENCE OF FORMULATIONS

The main curing ingredients are salt, nitrate and/or nitrite, glucose and ascorbic acid. The main effects of these agents on muscle enzymes is summarized in Tables 10.1 and 10.2, at concentrations usually found in the processing of dry-cured ham, and in Tables 10.3 and 10.4 in the case of dry-fermented sausage.

Salt is a traditional and very important curing agent and, in fact, strongly affects, positively or negatively, all the muscle enzymes. Cathepsins and aminopeptidases, except aminopeptidase B, are inhibited by salt, especially at high concentrations. Thus, the supplementation of an additional amount of salt to hams having very high cathepsin activity constitutes an easy way to prevent defects in the final texture, especially softness. In fact, softness has been found to be related to protein breakdown, which is also linked to higher residual cathepsin B activity and, to a lesser extent, to lower salt content (Parolari *et al.*

TABLE 10.1.
EFFECT OF TYPICAL CURING AGENTS CONCENTRATION AND OF PROCESSING CONDITIONS OF DRY-CURED HAM ON MUSCLE ENDOPROTEASES AND LIPASES[a]

Conditions[1]	Cathepsin B	Cathepsin B+L	Cathepsin H	Cathepsin D	m-calpain	Acid lipase	Neutral lipase	Acid esterase	Neutral esterase
NaCl added (6%)	⬂	⬂	⬂	⇩	⬂	⇧	⇩	⬂	⬂
NO_3^- added (200 ppm)	⇔	⇔	⇔	⇔	⇔	⇔	⇔	⇔	⇔
Ascorbic acid added (100 ppm)	⇔	⇔	⇔	⇔	⬂	⇔	⇔	⇔	⬂
Glucose added (200 ppm)	⬀	⇔	⇔	⇔	⬀	⇔	⇔	⇔	⇔
pH during process (6.0-6.3)	⇔	⇔	⇔	⬂	⬀	⬂	⬂	⬂	⇔
Drying temp. (20-23°C)	⇧	⇧	⇧	⇧	⇧	⇧	⬀	⬀	⬀
Ripening temp. (12-15°C)	⬀	⬀	⬀	⬀	⬀	⬀	⬀	⇔	⇔
Water activity (a_w = 0.90)	⬂	⇔	⬂	⬂	⇔	⬀	⬂	⇔	⬂

[a]Adapted from Rico *et al.* 1990, 1991, Toldrá 1992, Toldrá *et al.* 1992, 1993, Motilva *et al.* 1992, 1993, Rosell and Toldrá 1996

[1] ⇧ Strong activation, ⬀ moderate activation, ⇔ standard activity, ⬂ moderate inhibition, ⇩ strong inhibition

TABLE 10.2.
EFFECT OF TYPICAL CURING AGENTS CONCENTRATION AND OF PROCESSING CONDITIONS OF DRY-CURED HAM ON MUSCLE EXOPEPTIDASES[a]

Conditions[1]	Alanyl ap.	Arginyl ap.	Leucyl ap.	Pyro-glutamyl ap.	DPP I	DPP II	DPP III	DPP IV
NaCl added (6%)	⇘	⇗	⇔	⇩	⇔	⇘	⇘	⇘
NO_3^- added (200 ppm)	⇘	⇘	⇔	⇔	⇔	⇔	⇔	⇔
Ascorbic acid added (100 ppm)	⇘	⇔	⇔	⇘	-	-	-	-
Glucose added (200 ppm)	⇗	⇔	⇧	⇔	-	-	-	-
pH during processing (6.0-6.3)	⇔	⇔	⇘	⇘	⇧	⇧	⇩	⇘
Drying temperature (20-23°C)	⇧	⇧	⇧	⇧	⇧	⇧	⇧	⇧
Ripening temp. (12-15°C)	⇗	⇗	⇗	⇗	⇗	⇗	⇗	⇗
Water activity (a_w = 0.90)	⇘	⇘	⇘	⇘	⇘	⇘	⇘	⇘

[a]Adapted from Toldrá 1992, Toldrá *et al.* 1992b, 1993, Flores *et al.* 1997, Sentandreu and Toldrá 2001

[1] ⇧ Strong activation, ⇗ moderate activation, ⇔ standard activity, ⇘ moderate inhibition, ⇩ strong inhibition, - non available data

TABLE 10.3.
EFFECT OF TYPICAL CURING AGENTS CONCENTRATION AND OF PROCESSING CONDITIONS OF DRY-FERMENTED SAUSAGES ON MUSCLE ENDOPROTEASES AND LIPASES[a]

Conditions[1]	Cathepsin B	Cathepsin B+L	Cathepsin H	Cathepsin D	m-calpain	Acid lipase	Neutral lipase	Acid esterase	Neutral esterase
NaCl added (2%)	⬂	⬂	⬂	⇩	⬀	⬀	⬂	⇔	⇔
NO_3^- /NO_2^- added (250 ppm)	⬀	⇔	⇔	⬀	⇔	⇔	⇔	⇔	⇔
Ascorbic acid added (400 ppm)	⇔	⇔	⬂	⇔	⬂	⬂	⬂	⬂	⬂
Glucose added (2 g/L)	⬀	⬂	⬀	⬀	⬀	⇔	⇔	⇔	⇔
Fermentation pH (4.5-4.9)	⬂	⬂	⇩	⬀	⇩	⇧	⇩	⇧	⇩
Ferment. Temp. (23-25°C)	⇧	⇧	⇧	⇧	⇧	⇧	⬀	⬀	⬀
Drying/ripening pH (5.5-5.8)	⬀	⬀	⬀	⇔	⇔	⬀	⇩	⬀	⇩
Drying/ripening temp (10-15°C)	⬀	⬀	⬀	⬀	⬀	⬀	⬀	⇔	⇔
Water activity (a_w = 0.85-0.90)	⬂	⬂	⇩	⬂	⬂	⬀	⬂	⬂	⬂

[a]Adapted from Rico *et al.* 1990, 1991, Toldrá 1992, Toldrá *et al.* 1992, 1993, Motilva *et al.* 1992, 1993, Rosell and Toldrá 1996

[1] ⇧ Strong activation, ⬀ moderate activation, ⇔ standard activity, ⬂ moderate inhibition, ⇩ strong inhibition

TABLE 10.4.
EFFECT OF TYPICAL CURING AGENTS CONCENTRATION AND OF PROCESSING CONDITIONS OF DRY-FERMENTED SAUSAGES ON MUSCLE EXOPEPTIDASES[a]

Conditions[1]	Alanyl ap.	Arginyl ap.	Leucyl ap.	Pyro-glutamyl ap.	DPP I	DPP II	DPP III	DPP IV
NaCl added (2%)	⬂	⬀	⇔	⇔	⇔	⬂	⬂	⬂
NO_3^- /NO_2^- added (250 ppm)	⬂	⬂	⇔	⇔	⇔	⇔	⇔	⇔
Ascorbic acid added (400 ppm)	⬂	⇔	⬂	⇔	-	-	-	-
Glucose added (2 g/L)	⇔	⇔	⇧	⇔	-	-	-	-
Fermentation pH (4.5-4.9)	⇩	⇩	⬂	⇩	⇧	⇧	⇩	⇩
Fermentation temp (23-25°C)	⇧	⇧	⇧	⇧	⇧	⇧	⇧	⇧
Drying/ripening pH (5.5-5.8)	⬂	⬂	⬂	⇩	⇧	⇧	⇩	⇩
Drying/ripening temp (10-15°C)	⬀	⬀	⬀	⬀	⬀	⬀	⬀	⬀
Water activity (a_w = 0.85-0.90)	⬂	⬂	⬂	⬂	⬂	⬂	⬂	⬂

[a]Adapted from Toldrá 1992, Toldrá *et al.* 1992b, 1993, Flores *et al.* 1997, Sentandreu and Toldrá 2001

[1] ⇧ Strong activation, ⬀ moderate activation, ⇔ standard activity, ⬂ moderate inhibition, ⇩ strong inhibition, - non available data

1994). However, there are two disadvantages when adding an excess of salt: the negative impact on the nutritional quality because the final product would contain too much salt and the sensory quality, which is affected by an excessive salty taste.

Muscle neutral lipase, neutral esterase and acid esterase are inhibited by salt, whereas acid lipase is activated. Arginyl aminopeptidase (or aminopeptidase B) is a chloride-activated enzyme and its activity is enhanced up to three times at 0.35M NaCl. Something similar happens with m-calpain, which is also activated below 0.5M NaCl in the case of dry-fermented sausages but is inhibited at higher concentrations like those found in dry-cured ham. Nitrates and nitrites do not exert a noticeable effect on the enzyme activities. Ascorbic acid at high concentrations, such as those found in dry-fermented sausages, exerts a slight inhibitory effect on cathepsin H, m-calpain, leucyl aminopeptidase, lipases and esterases. Glucose has a positive effect on some enzymes like cathepsin B, m-calpain, alanyl and leucyl aminopeptidases. Cathepsin D and H would be also activated by glucose at the concentrations found during the processing of dry-fermented sausages.

IMPORTANCE OF PROCESSING CONDITIONS

Effect of Prefreezing/Thawing of Raw Hams Prior to Dry-Curing

Prefreezing of hams produces several effects on quality attributes of dry-cured products although it is not allowed for processing of Parma and Bayonne hams because freezing is considered detrimental to the sensory quality of the final product. The frozen storage of hams in standard production is common in situations of over supply and lower prices. It has the advantage to increase in porosity upon thawing, which results in a better penetration of the curing ingredients. However, salt penetration is quicker, and salting time must be reduced to avoid an excessive salt uptake and the consequent salty taste at the end of the process. A summary of main changes between standard hams and frozen/thawed hams are shown in Table 10.5.

The water loss after curing and after aging in the prefrozen hams is greater than that reported for unfrozen hams. Prefrozen/thawed hams experience more intense biochemical changes, especially in the first months of process, although these changes are not enough to have a noticeable effect on the final sensory quality of the hams. Therefore, the lipolysis is more intense between 0-5 months in muscle and 0-10 months in adipose tissue, reaching similar values at the end of the process (Motilva *et al.* 1994). It must be taken into account that depending on the frozen storage conditions (time and temperature), lipases may be more active and increase the initial levels of free fatty acids. The prolonged period of frozen storage may adversely affect the final flavor of the hams (Graham and Blumer 1972).

TABLE 10.5.
CHARACTERISTICS OF DRY-CURED HAMS: EXAMPLE OF THE EFFECT OF PRE-FREEZING AND THAWING OF HAMS IN COMPARISON TO STANDARD HAMS[a]

Mean values (25 hams)	**Standard hams**	**Prefrozen/thawed hams**
Initial mean weight (Kg)	9.0	9.0
Processing time (months)	15	15
Final product :		
Protein (g/100g)	31.5	32.6
Lipids (g/100g)	2.7	2.8
Moisture (g/100g)	59	56
NaCl (%)	5.3	7.1
Weight loss (%)	35	37
pH	6.1	6.2
a_w	0.84	0.83
Free fatty acids (g/100 g fat)	13.1	13.2
Carbonyl index (μmol CO/g DM)	2.18	1.45

[a]Adapted from Motilva *et al.* 1994

The proteolysis levels are also higher in the zones of the ham where water losses and absorption of salt are slowest, like the muscle *Biceps femoris*. This is evident because a high incidence of white precipitates (tyrosine crystals) is usually found in that muscle (Bañón *et al.* 1999). Nucleation and growth are involved in the formation of these crystals. Nucleation is favored by freezing/thawing because cell membranes may be disrupted and act as heterogeneous nucleation sites (Arnau *et al.* 1994). In unfrozen hams, this nucleation and/or diffusion is slower and reduces the formation of tyrosine crystals during drying/ripening. However, this increased lipolysis and proteolysis has not been reported to affect the final sensory quality except for a higher salty taste (Motilva *et al.* 1994; Bañón *et al.* 1999). Similar sensory scores and microbial counts have been observed in frozen hams cured without thawing (Kemp *et al.* 1982).

Effect of Processing Conditions

The effect of processing conditions on muscle enzymes performed in meat model systems representing dry-cured ham is shown in Tables 10.1 and 10.2. In the case of dry-fermented sausages, the effect of processing conditions is summarized in Tables 10.3 and 10.4. The dry-curing process can be considered, in general, a mild process since temperatures do not usually exceed 30C. The temperature during the salting and post-salting stages is kept below 6C to retard the enzyme action. Lipases and esterases are not affected as proteases are at low temperatures, and thus, a low enzyme activity can be expected even at refrigeration temperatures such as those found during the salting and post-salting stages. The use of alternating aging temperatures in the manufacture of country-style hams has been reported to produce the development of free fatty acids at a faster rate (Kemp *et al.* 1968). However, the more intense biochemical changes become apparent during the drying (in the case of hams) or fermentation (in the case of sausages) stages where the temperature rises up to 20–25C. Noticeable changes are also observed at milder temperatures such as 14-16C during ripening for several weeks (for sausages) or months (for hams).

The pH is quite constant along the processing of dry-cured ham, starting at 5.5-5.8 and finishing at 6.3-6.6 at the end of the process. Most of the muscle enzymes are quite active at this pH interval. The exception is cathepsin D, an acid proteinase that shows low activity at a pH higher than 6.0. The pH found in ham along the process is not far from the optimal activity of acid lipase and esterase, although neutral lipases and esterases would show very low activity. A reverse fact can be found in the case of dry-fermented sausages where the acid pH (below 5.0) favors the activity of acid enzymes such as cathepsin D and acid lipase and esterase. The negative redox potential (between –150 mV and – 250 mV) found during the dry-curing process favors the activity of cathepsins and some aminopeptidases due to the reducing conditions they need for optimal activity (Motilva *et al.* 1993). Water activity into the products is also important in controlling the enzyme activity, especially after midprocess or towards the end when a_w is reduced to values below 0.90. Muscle enzymes, with the exception of m-calpain and acid lipase, are strongly affected by the decrease in a_w along the processing of dry-cured meat products.

The length of the process is also of essential importance for the activity of the enzymes. It is evident that a longer processing time allows a more pronounced action of the enzymes. A dry-fermented sausage will experience a more intense proteolysis and lipolysis when the ripening time is longer. Most of the enzymes are stable for the whole process and even up to two years, as is the case of some hams (Toldrá and Flores 1998). The apparent half-life (time necessary for the reduction of the enzyme activity to half of the initial activity) has been calculated for most of the muscle enzymes based on its activity during different dry-curing processes (Table 10.6). It can be observed that, in general, the stability is very good and allows for the enzyme action during the full process. Only the action of m-calpain and pyroglutamylaminopeptidase would be reduced to the initial weeks of process in hams or initial days in sausages.

TABLE 10.6.
APPROXIMATED STABILITY OF PORK MUSCLE ENZYMES. DATA EXPRESSED AS AN APPARENT HALF-LIFE ($T_{1/2}$ IN MONTHS) THAT GIVES AN ESTIMATION OF THE TIME NECESSARY TO REDUCE 50% OF THE INITIAL ACTIVITY. DATA ESTIMATED FROM THE IBERIAN AND SERRANO DRY-CURING PROCESSES

Endopeptidases	$t_{1/2}$	Exopeptidases	$t_{1/2}$	Lipolytic enzymes	$t_{1/2}$
Cathepsin B	5	Alanyl aminop.	4	Acid lipase	4
Cathepsin B+L	5	Arginyl aminop.	2.5	Neutral lipase	>10
Cathepsin H	2.5	Leucyl aminop.	9	Phospholipase A	4
Cathepsin D	2.5	Pyroglutamyl	1	Acid esterase	4
m-calpain	0.3	Dipeptidyl pep I	3.5	Neutral esterase	7
μ-calpain	<0.01	Dipeptidyl pep II	2.5	Adipose neutral lip.	2.5
		Dipeptidyl pep III	>10	Adipose basic lipase	2.5
		Dipeptidyl pep IV	2	Adipose acid esterase	7
		Tripeptidyl pep	5	Adipose neutral est.	>10

(Adapted from Toldrá and Flores 1998)

Effect of Accelerated Processing

There is a trend, mainly for economical reasons, to reduce the time for processing without affecting quality or safety. Some attempts to reduce time of ham processing have been performed, like bacterial inoculation, enzymes addition, etc., but with some negative effects on sensory quality (see Chap. 3). The tumbling has received better results for acceleration and uniformity in salt distribution, although with some negative aspects such as physical damage to the hams. In the case of dry-fermented sausages (as discussed in Chap. 4), the addition of starter cultures has given important positive contributions to safety, standardization, reduction in processing time and, sometimes, improvement in sensory quality. However, the addition of specific enzymes has not been so satisfactory, giving more variability in the final quality.

REFERENCES

ARMERO, E., BARBOSA, J.A., TOLDRÁ, F., BASELGA, M. and PLA, M. 1999a. Effects of the terminal sire and sex on pork muscle cathepsin (B, B+L and H), cysteine proteinase inhibitors and lipolytic enzyme activities. Meat Sci. *51*, 185–189.

ARMERO, E., BASELGA, M., ARISTOY, M-C. and TOLDRÁ, F. 1999b. Effects of sire type and sex on pork muscle exopeptidase activity, natural dipeptides and free amino acids. J. Sci. Food Agric. *79*, 1280–1284.

ARMERO, E., FLORES, M., TOLDRÁ, F., BARBOSA, J.A., OLIVET, J., PLA, M. and BASELGA, M. 1999c. Effects of pig sire type and sex on carcass traits, meat quality and sensory quality of dry-cured ham. J. Sci. Food Agric. *79*, 1147–1154.

ARMERO, E., NAVARRO, J.L., NADAL, M.I., BASELGA, M. and TOLDRÁ, F. 2002. Lipid composition of pork muscle as affected by sire genetic type. J. Food Biochem. *26*, 91–102.

ARNAU, J., GOU, P. and GUERRERO, L. 1994. The effects of freezing, meat pH and storage temperature on the formation of white film and tyrosine crystals in dry-cured hams. J. Sci. Food Agric. *66*, 279–282.

ASHBY, B.H., JAMES, G.M. and KRAMER, A. 1973. Effects of freezing and packaging methods on shrinkage of hams in frozen storage. J. Food Sci. *38*, 254–257.

BAÑÓN, S., CAYUELA, J.M., GRANADOS, M.V. and GARRIDO, M.D. 1999. Pre-cure freezing affects proteolysis in dry-cured hams. Meat Sci. *51*, 11–16.

BERDAGUÉ, J.L., BONNAUD, N., ROUSSET, S. and TOURAILLE, C. 1993. Influence of pig crossbreed on the composition, volatile compound content and flavour of dry cured ham. Meat Sci. *34*, 119–129.

BLASCO, A., GOU, P., GISPERT, M., ESTANY, J., SOLER, Q., DIESTRE, A. and TIBAU, J. 1994. Comparison of five types of pig crosses. I. growth and carcass traits. Livestock Prod. Sci. *40*, 171–178.

DE LA HOZ, L., LOPEZ, M.O., HIERRO, E., CAMBERO, M.I. and ORDÓÑEZ, J.A. 1996. Effect of diet on the fatty acid composition of intramuscular and intermuscular fat in Iberian pig cured hams. Food Sci. Technol. Int. *2*, 391–397.

ENSER, M., WOOD, J.D., MONCRIEFF, C.B. and KEMPSTER, A.J. 1988. Chemical composition of backfat in 58 Kg pig carcasses covering a wide range of fat thickness. Anim. Prod. *46*, 503–504.

FLORES, J., BIRON, C., IZQUIERDO, L. and NIETO, P. 1988. Characterization of green hams from Iberian pigs by fast analysis of subcutaneous fat. Meat Sci. *23*, 253–262.

FLORES, M., ROMERO, J., ARISTOY, M-C., FLORES, J. and TOLDRÁ, F. 1994. Differences in muscle proteolytic activities among pork breed types. Sci. des Aliment. *14*, 469–474.

FLORES, M., ARISTOY, M-C. and TOLDRÁ, F. 1997. Curing agents affect aminopeptidase activity from porcine skeletal muscle. Z. Lebensm. Unters Forsch. *205*, 343–346.

GALLO, L., MONTOBBIO, P., CARNIER, P. and BITTANTE, G. 1994. Breed and crossbreeding effects on weight, yield and quality of heavy Italian dry-cured hams. Livestock Prod. Sci. *40*, 197–205.

GOU, P., GUERRERO, L. and ARNAU, J. 1995. Sex and crossbreed effects on the characteristics of dry-cured ham. Meat Sci. *40*, 21–31.

GRAHAM, P.P. and BLUMER, T.N. 1972. Effect of pre-freezing hams on quality attributes of dry-cured product. J. Milk Food Technol. *35*, 98–101.

GUERRERO, L., GOU, P., ALONSO, P. and ARNAU, J. 1996. Study of the physicochemical and sensorial characteristics of dry-cured hams in three pig genetic types. J. Sci. Food Agric. *70*, 526–530.

ISABEL, B., LÓPEZ-BOTE, C.J., REY, A.I. and SANZ, R. 1999. Influence of dietary α-tocopheryl acetate supplementation of pigs on oxidative deterioration and weight loss in sliced dry-cured ham. Meat Sci. *51*, 227–232.

KEMP, J.D., SMITH, R.H. and MOODY, W.G. 1968. Quality of aged hams as affected by alternating aging temperatures. Food Tech. *22*, 113–114.

KEMP, J.D., LANGLOIS, B.E. and JOHNSON, A.E. 1982. Effect of pre-cure freezing and thawing on the microflora, fat characteristics and palatability of dry-cured ham. J. Food Protect. *45*, 244–248.

LARICK, D.K., TURNER, B.E., SCHOENHERR, W.D., COFFEY, M.T. and PILKINGTON, D.H. 1992. Volatile compound content and fatty acid composition of pork as influenced by linoleic acid content of the diet. J. Anim. Sci. *70*, 1397–1403.

MONAHAN, F.J., BUCKLEY, D.J., MORRISEY, P.A., LYNCH, P.B. and GRAY, J.I. 1992. Influence of dietary fat and α-tocopherol supplementation on lipid oxidation in pork. Meat Sci. *31*, 229–241.

MORGAN, C.A., NOBLE, R.C., COCCHI, M. and MCCARTNEY, R. 1992. Manipulation of the fatty acid composition of pig meat lipid by dietary means. J. Sci. Food Agric. *58*, 357–368.

MOTILVA, M-J., RICO, E. and TOLDRÁ, F. 1993. Effect of redox potential on the muscle enzyme system. J. Food Biochem. *16*, 207–215.

MOTILVA, M-J., TOLDRÁ, F. and FLORES, J. 1992. Assay of lipase and esterase activities in fresh pork meat and dry-cured ham. Z. Lebens Unters. Forsch. *195*, 446–450.

MOTILVA, M-J. and TOLDRÁ, F. 1993. Effect of curing agents and water activity on pork muscle and adipose subcutaneous tissue lipolytic activity. Z. Lebens Unters. Forsch. *196*, 228–232.

MOTILVA, M-J., TOLDRÁ, F., NADAL, M-I. and FLORES, J. 1994. Prefreezing of hams affects lipolysis during dry-curing. J. Food Sci. *59*, 303–305.

OLIVER, M.A., GOU, P., GISPERT, M., DIESTRE, A., ARNAU, J., NOGUERA, J.L. and BLASCO, A. 1994. Comparison of five types of pig crosses. II. fresh meat quality and sensory characteristics of dry-cured ham. Livestock Prod. Sci. *40*, 179–185.

PAROLARI, G., VIRGILI, R. and SCHIVAZZAPPA, C. 1994. Relationship between cathepsin B activity and compositional parameters in dry-cured hams of normal and defective texture. Meat Sci. *38*, 117–122.

RHEE, K.S. 1992. Fatty acids in meat and meat products. *In* Fatty Acids in Foods and Their Health Implications. C. K. Chow (ed.), p. 65. Marcel Dekker, New York.

RICO, E., TOLDRÁ, F. and FLORES, J. 1990. Activity of cathepsin D as affected by chemical and physical dry-curing parameters. Z. Lebensm. Unters. Forsch. *191*, 20–23.

RICO, E., TOLDRÁ, F. and FLORES, J. 1991. Effect of dry-curing process parameters on pork muscle cathepsins B, H and L activities. Z. Lebensm. Unters. Forsch. *193*, 541–544.

ROSELL, C-M. and TOLDRÁ, F. 1996. Effect of curing agents on m-calpain activity through the curing process. Z. Lebensm. Unters. Forchs. *203*, 320–325.

ROSELL, C-M. and TOLDRÁ, F. 1998. Comparison of muscle proteolytic and lipolytic enzyme levels in raw hams from Iberian and White pigs. J. Sci. Food Agric. *76*, 117–122.

RUIZ, J., CAVA, R., ANTEQUERA, T., MARTÍN, L., VENTANAS, J. and LÓPEZ-BOTE, C.L. 1998. Prediction of the feeding background of Iberian pigs using the fatty acid profile of subcutaneous, muscle and hepatic fat. Meat Sci. *49*, 155–163.

SÁRRAGA, C., GIL, M. and GARCÍA-REGUEIRO, J.A. 1993. Comparison of calpain and cathepsin (B, L and D) activities during dry-cured ham processing from heavy and light large white pigs. J. Food Sci. Agric. *62*, 71–75.

SCHIVAZZAPPA, C., VIRGILI, R., DEGNI, M. and CERATI, C. 1998. Effects of different pig lines on some characteristics of Parma ham. Ind. Conserve *73*, 110–116.

SENTANDREU, M.A. and TOLDRA, F. 2001. Dipeptidyl peptidase activities along the processing of Serrano dry-cured ham. Eur. Food Res. Technol. *213*, 83–87.

TOLDRÁ, F. 1992. The enzymology of dry-curing of meat products. *In* New Technologies for Meat and Meat Products. F.J.M. Smulders, F. Toldrá, J. Flores and M. Prieto (eds.), pp. 209–231. Audet, Nijmegen, The Netherlands.

TOLDRÁ, F., RICO, E. and FLORES, J. 1992. Activities of pork muscle proteases in cured meats. Biochimie *74*, 291-296.

TOLDRÁ, F., CERVERÓ, M-C. and PART, C. 1993. Porcine aminopeptidase activity as affected by curing agents. J. Food Sci. *58*, 724–726, 747.

TOLDRÁ, F., FLORES, M., ARISTOY, M-C., VIRGILI, R. and PAROLARI, G. 1996a. Pattern of muscle proteolytic and lipolytic enzymes from light and heavy pigs. J. Sci. Food Agric. *71*, 124–128.

TOLDRÁ, F., REIG, M., HERNÁNDEZ, P. and NAVARRO, J.L. 1996b. Lipids from pork meat as related to a healthy diet. Recent Res. Devel. In Nutrition *1*, 79–86.

CHAPTER 11

MAIN DEFECTS AND PREVENTIVE MEASURES

The main defects in dry-cured meat products are mainly due to physicochemical changes or the development of undesirable microorganisms either in the raw materials or during the processing. The defects are thus grouped into one of these categories, although they rarely appear alone. These defects are related to several sensory characteristics like flavor, texture, color and appearance. It must be taken into account that dry-curing processes are rather complex and a complete and systematic control of the full process is sometimes necessary in order to find the specific cause for a particular defect. In this chapter, the probable causes and consequences as well as preventive measures are given for each defect.

MAIN DEFECTS IN DRY-CURED HAMS AND THEIR CAUSES

Physicochemical Change

Boar Taint. The development of a strong and unpleasant odor called boar taint can be detected, especially in fatty tissue of hams from sexually mature intact male pigs (raw material defect). The main compound responsible for this off-flavor is the male hormone 5-α-androsten-16-en-3-one (androstenone), which has an intense odor of urine. The odor is especially detected in hams from noncastrated old boars but is not as common in boars slaughtered at light weights. This is a fat-soluble hormone that is accumulated in fat and is noticeable when it is accumulated above its sensory threshold value, which is around 0.5 μg/g. This problem is caused by the higher yields in lean meat and the feed efficiency of young males that have not been castrated. Other compounds like indol and skatole, produced from the amino acid tryptophan in the gut, may also contribute to boar taint. If there is uncertainty about the castration of male pigs, some industries prefer to process hams from female pigs in order to avoid potential off-flavor problems.

Salty Taste. An excessive salty taste can develop that is mainly due to the presence of an excess of salt in the ham (processing technology defect). This can be caused by an excessive supply of salt during the salting stage. This is common in either PSE or frozen/thawed hams. Because salt uptake is related to drip loss, PSE hams and frozen/thawed hams absorb a higher amount of salt than normal hams (Arnau *et al.* 1995). In these cases, it is necessary to control the salt uptake by reducing the duration and/or amount of added salt in the salting stage.

Bitter Taste. A bitter-like taste is associated with a high proteolysis index above 35% (Careri *et al.* 1993) with a sensible accumulation of bitter peptides and hydrophobic amino acids generated from an intensive protein breakdown by muscle proteases (processing technology defect). Bitterness is related to the average hydrophobicity of peptides and free amino acids (Habibi-Najafi and Lee 1996). This is given by the Q value ($Q = \Sigma\Delta f/n$) expressed as the relation between the addition of the hydrophobicity values for all amino acids in the peptide ($\Sigma\Delta f$) to the number of residues (n). So, bitterness would be found in peptides with Q values above 1400 (Ney 1979). This is the case of some dipeptides, like Ile-Val, Leu-Glu, Ile-Asp and Pro-Leu, found in dry-cured ham (Sentandreu *et al.* 2002).

Sheepy or Fishy Aromas. Sheep- or fish-like aromas can develop, especially in fat tissue, during ripening. The origin of these off-flavors is the use of feeds with the presence of sheep or fish meals in places where the addition of these meals is allowed (raw materials defect). If they are not correctly deodorized, the aromas may accumulate in the fat tissue of pigs and, because their thresholds are very low (in the case of trimethylamine in fish), the off-flavors may be perceived by consumers. In these cases, the control of feed is essential for prevention.

Rancidity. Rancid off-flavors and yellowish color in the fat tissue can develop as a result of the oxidation of unsaturated free fatty acids. The main cause of rancidity is the use of feeds with fats having high levels of unsaturation or already oxidized fatty acids (raw materials/processing technology defect). Pigs are monogastric and directly use the fatty acids provided in the feed. So, the accumulation of high levels of unsaturated fatty acids constitutes a risk for oxidation, especially if hams are ripened at relatively high temperatures or for a long period of time.

Lack of Flavor. Very poor development of flavor can often be found in hams with very short processing times (processing technology defect). Insufficient time for completion of biochemical reactions, especially proteolysis and lipolysis, and thus poor production of flavor precursors and flavor compounds, or an excessive pH drop that inhibits most of the biochemical reactions responsible for flavor development or a poor enzymatic profile.

Hardening. This refers to the texture of the ham being too hard. Excessive drying as a result of the use of high temperature, low relative humidity and/or high air velocity (processing technology defect) can cause hardening. Sometimes, an associated fat exudation is also observed. In extreme cases, muscle disjunction may appear, leaving holes or fissures where mites, molds or other microorganisms could penetrate. Hardness may be checked by hand compression of lean meat. As an effective protective measure and once the desired water loss is reached, hams are usually protected from further drying with a coating of food-grade fat.

Softness. Hams present a soft, sometimes gummy texture, especially in inner muscles *Biceps femoris* and *Semitendinosus*. There are several causes such as an excess of proteolysis, especially by muscle endoproteases, producing a great destruction of the myofibrillar protein structure (raw materials/processing technology defect). The solution is to check for cathepsin B activity in raw hams and add an excess of salt to control protein breakdown in those hams with high proteolytic activity (Virgili *et al.* 1995). Another cause is an incomplete water equalization or limited fiber swelling because of inefficient drying (i.e., high relative humidity) that results in poor water loss and an excess of moisture inside the ham or a very fast drying (high temperature, low relative humidity and high air velocity) that produces hardening of the outer ham surface, retaining most of the moisture inside the ham.

White Crystals Inside the Ham

The formation of small white spots in the ham can appear randomly distributed through the whole piece, especially in the zones with higher moisture content (processing technology defect). These crystals are produced as a consequence of an excessive proteolysis, especially for the combined action of cathepsins and muscle aminopeptidases that generate free amino acids. The white crystals are mainly composed of tyrosine, and the formation depends on its concentration and solubility at the pH and temperature of the ham (Arnau *et al.* 1996). White crystals are typically found in frozen/thawed hams that are aged for long periods of time, where the structural changes can affect nucleation and crystal growth.

White Film in Vacuum-Packed Ham

A white film can form on the cut surface of vacuum-packed ham slices. This process is very fast, becoming apparent in a few days of refrigerated storage, especially on the external surface in contact with the plastic bag (processing technology defect). The film develops mainly on the *Semimembranosus*, *Semitendinosus* and *Biceps femoris* muscles. The causes have been studied, but the microbial film as a cause has been discounted (Butz *et al.* 1974; Cantoni *et al.* 1987). The structure of the film has been observed by scanning electron microscopy, revealing a foam-like fibrous structure composed of protein materials and the absence of microorganisms. Its composition in proteins and in free amino acids, analyzed by electrophoresis and reverse-phase HPLC, respectively (Toldrá *et al.* 1990), revealed a major presence of sarcoplasmic and low molecular weight myofibrillar proteins, a low amount (below 12%) of free amino acids but noticeable tyrosine. Some factors, such as high moisture and low chloride content combined with a high vacuum degree, favor the formation of this film.

Yellow Color in Fat. A yellowish color in the fat tissue can develop, resulting from the oxidation of unsaturated free fatty acids. The main reason for this is the use of feeds with fats having a high level of unsaturation or already oxidized fatty acids (raw materials defect). Unsaturated fatty acids risk oxidation, especially if hams are ripened at relatively high temperatures or for a long period of time.

Deficient Curing Coloration. The correct cured color is not formed in specific areas because the reduction of nitrate to nitrite is not achieved, or myoglobin protein is abnormally cleaved by muscle endoproteases (processing technology defect). In other cases, small brown blood spots have some occurrence and are due to capillary hemorrhages from electrical or, to a lesser extent, CO_2 stunning of pigs (Parolari 1996).

Development of Undesirable Microorganisms

These defects were more important a few decades ago, but the availability of good refrigeration systems in most industries and knowledge of limiting salt and water activity necessary to inhibit growth of spoilage microorganisms have reduced its relative importance in recent years. In general, off-odors resulting from microbial spoilage may be assessed by the probe-and-sniff technique, consisting of the quick insertion of a probe, a thin horse bone, near the aitch bone followed by immediate sniffing. The main defects are listed below.

Sourness. A sour taste is developed near the bones as a result of anaerobic bacterial production and subsequent accumulation of short-chain fatty acids and lactic acid (raw materials/processing technology defect). It may be due to the growth of lactic acid bacteria. It is usually a problem in the cold chain or nonhomogeneous salt diffusion or in openings between muscles.

Putrefaction. The development of hydrogen sulphide, methanethiol, ammonia, indol, etc., can give off putrefactive odors (raw materials/processing technology defect). It is usually accompanied by green or grey colors, resulting from muscle pigment degradation. The development of odors may be due to the growth of putrefactive, facultative anaerobic strains. The origin may be a problem of hygiene, contamination of the carcass at the slaughterhouse, hams with high pH (DFD hams) prone to contamination, incorrect refrigeration during salting and/or nonhomogeneous salt diffusion or incorrect drying that leaves some areas with high moisture content. Preventive measures must control the raw material, which must be hygienic and correctly cold, avoid hams having high pH and control refrigeration temperatures during transportation, reception and salting stages.

Phenic Acid. A phenol-like off-odor can develop during the ripening/drying process (processing technology defect). This type of odor is due to mold spoilage from the genus Penicillium, mostly *P. commune*. Preventive measures would include controlling the surface contamination to levels lower than 10^2 spores, adding salt on the aitch bone to inactivate spores potentially present and preventing penetration of *P. commune* during ripening (Spotti *et al.* 1988).

Potato Aroma. An odor similar to that of potato or earth can develop (raw materials/processing technology defect). Pyrazines, especially mono-, di- and tri-methylpyrazines, indicate the appearance of this off-flavor caused by the 2-methoxy-3-isopropylpyrazine, the impact compound of potato odor in spoiled hams. Apparently, the main culprit responsible for this off-flavor is *Pseudomonas cepacia*, a bacteria isolated from the aitch bone (Blanco *et al.* 1994). Preventive measures consist of the same measures mentioned for phenic acid.

Presence of Molds. Mold growth can occur inside, especially around the aitch bone or on the outer surface (processing technology defect). The genus *Penicillium* may grow during the initial stages because it grows well at low temperatures, while *Aspergillus* develop later because they prefer higher temperatures such as those found during drying. The color (green, black, violet, etc.) will also vary, depending on the type of mold growing on the surface.

The production of mycotoxins depends on the nutritive media, pH, a_w, temperature and time of storage (Arnau 1988). Preventive measures consist of controlling surface contamination by smoking or immersion in sorbate solutions, adding salt on the aitch bone to inactivate spores potentially present and preventing the penetration of molds during ripening.

Abnormal Coloration. Some discoloration may appear in certain areas of the ham, probably due to the growth of peroxide-forming lactic acid bacteria (processing technology defect).

Other Defects

Mites Growth. Mites (accari) may appear during the drying/aging stages and may produce small holes inside the ham after long ripening times (processing technology defect). Another important risk for plant operators and consumers consists of dermal and/or respiratory allergy caused by mites. The genera Tyrophagus, one of the most extended, grows well at temperatures between 11C and 37C, relative humidities above 65–70% and scarce light (Lorenzo and Flores 1988). Mites are small with a white color and relatively rapid movement. They are first detected by their accumulation outside the ham and when they grow too much and fall on the floor forming small mounds. Preventive measures consist of a strict hygienic control of the chambers, hams and personnel; periodical exhaustive disinfection; and control of relative humidity and temperature during aging. Hams may be successfully protected by covering with food-grade fat, like lard, either manually rubbed on the surface or by subtle immersion in melted fat at 60–65C.

MAIN DEFECTS IN DRY-FERMENTED SAUSAGES AND THEIR CAUSES

Physicochemical Changes

Sourness. A sour taste is developed as a result of an intense fermentation (processing technology defect). This sour taste is more frequent in sausages with larger diameters. There are several causes for sour taste: the use of a high temperature for fermentation that facilitates lactobacilli growth and lactic acid generation, an excessive amount of sugars added to the mix that allows higher lactic acid production and accumulation and, although in a minor scale, an intense lipolysis resulting in accumulation of free fatty acids. Preventive measures consist of the reduction of fermentation temperature if it is too high, control of sugar addition and type of sugar (glucose or saccharose are easily fermented by LAB) and avoidance of excessive drying (hardening) of external surfaces during ripening.

Rancidity. Rancid off-flavors and yellowish color in the fat tissue can develop from the oxidation of unsaturated free fatty acids (raw materials defect). This is caused by the use of lard with a high level of unsaturation or lard that contains oxidized fatty acids after being stored for a prolonged time prior to its use. Some bacterial strains may also produce peroxides that accelerate oxidative processes. Lard must be controlled before use, rejecting those fats either too unsaturated or stored for long periods of time. It is also important to exclude oxygen from the mixture, using mixing and stuffing under vacuum, and use starter culture strains with catalase activity.

Boar Taint. Unpleasant urine-like off-flavor can develop due to the accumulation of the male hormone 5-α-androsten-16-en-3-one (androstenone). As in the case of hams, this is due to the use of lard from noncastrated male pigs (raw materials defect).

Case Hardened. A hardened outer surface or crust can develop (processing technology defect). In the case of sausages of short caliber, this can occur due to excessive drying as a result of using high temperature, low relative humidity and/or high air velocity. In the case of sausages with larger diameters, it is common to find a hard dry area on the outer sausage surface that acts as a shield and retains most of the moisture, thus leaving softness inside the sausage. Low fat content may also contribute to the sausage hardness. The cross section shows a contrast of colors, dark red color on the outer surface and clear pink in the center.

Softness. Sausages present a soft texture, especially in the center (processing technology defect). There are several causes of softness, such as the application of inefficient drying (i.e., high relative humidity) that results in poor water loss and an excess of moisture inside the sausage, fat smearing that retains water, insufficient pH drop, an excess of fat in the mix or a very fast drying (high temperature, low relative humidity and high air velocity) that produces hardening of the outer sausage surface but retains most of the moisture inside the sausage. Fermentation (correct sugar addition and starter cultures function) and drying must be carefully controlled, especially relative humidity and air velocity.

Wrinkled Casing. The casing is not retracted together with the sausage mix during the ripening/drying (processing technology defect). The casing gets detached and wrinkled as a result of a very fast drying.

Holes and/or Fissures. Sausages present inner holes, sometimes relatively big, or fissures that give an unpleasant aspect to the product (raw materials/processing technology defect). The causes may proceed from a contaminated raw material, an excessive production of gas by heterofermentative bacteria (especially if high amounts of sugars are added) and/or a low pressure during stuffing. The excessive gas production not only can cause texture problems and pinholes but also even breakage of the casing due to the product expansion (Bacus 1984).

Unclear Cross Section. Pieces of fat are not clearly separated from lean meat (processing technology defect). This can occur due to an excess of moisture, unsaturated fat with low melting point or use of a low pressure during stuffing of the mix into the casing.

Yellow Color in Fat. A yellowish color in the fat tissue can develop, resulting from the oxidation of unsaturated free fatty acids which are usually associated with rancid aromas (raw materials defect). The main reason this happens is due to the use of fats with a high percentage of unsaturation.

Deficient Curing Coloration. The pH drop may be too fast and inhibit the activity of the nitrate-reductase enzyme, and thus, reduction of nitrate to nitrite is not achieved (processing technology defect). A pale color may be due to PSE pork meats or myoglobin protein cleavage by muscle endoproteases.

Other Abnormal Colors. There are several alterations of colors produced by different factors (processing technology defect). Color fading in the center of the sausage and darker color in the outside surface may appear when internal moisture is too high. An excess of acid production in those sausages with an intense pH drop may deteriorate color.

Development of Undesirable Microorganisms

Putrefaction. Putrefactive odors can develop from facultative anaerobic bacteria (raw materials/processing technology defect). These odors are usually accompanied by green or grey colors that result from muscle pigment degradation and gas generation. There are several possible causes: lack of hygiene, contamination of the carcass at the slaughterhouse, use of meats with high pH (DFD) prone to contamination, incorrect refrigeration during chopping and mixing, incorrect fermentation with insufficient pH drop or incorrect drying, which leaves some areas with high moisture content. Preventive measures must keep correct hygiene and control in all processing stages, especially the raw materials, including the correct management of meats and additives under refrigeration.

Molds. The presence of molds on the outer surface, which are desirable for some sausages because they contribute to final flavor, are however undesirable or rejected by consumers in other types of sausages (processing technology defect). Molds grow easily during ripening/drying, but the wrong molds may give unsatisfactory quality or may even produce mycotoxins. Molds may be avoided by dipping the sausages into a 2.5% potassium sorbate solutions or, once they have grown, may be brushed from the casing. Smoking also contributes some antifungal properties.

Abnormal Colors. Several alterations of colors or some discoloration may appear from peroxide-forming bacterial strains (processing technology defect).

REFERENCES

ARNAU, J., GUERRERO, L., HORTOS, M. and GARCIA-REGUEIRO, J.A. 1996. The composition of white film and white crystals found in dry-cured hams. J. Sci. Food Agric. *70*, 449–452.

ARNAU, J., GUERRERO, L., CASADEMONT, G. and GOU, P. 1995. Physical and chemical changes in different zones of normal and PSE dry cured ham during processing. Food Chem. *52*, 63–69.

ARNAU, J. 1988. Aparición de hongos en el interior y exterior del jamón. *In* Jamón Curado: Aspectos Técnicos. J. Arnau, M. Hugas and J.M. Monfort (eds.), pp. 171–176. Grafis-Sant, Girona, Spain.

BACUS, J. 1984. Update: Meat fermentation 1984. Food Technol. *38*, 58–63.

BLANCO, D., BARBIERI, G., MAMBRIANI, P., SPOTTI, E. and BARBUTI, S. 1994 Study of the potato defect of raw dry-cured ham. Ind. Conserve *69*, 230–236.

BUTZ, R.G., BLUMER, T.N., CHRISTIAN, J.A. and SWAISGOOD, H.E. 1974. Factors responsible for white film formation on cut surfaces of dry-cured hams. J. Food Sci. *39*, 516–519.

CANTONI, C., CATTANEO, P. and BRENNA, O. 1987. Una nuova alterazione del prociutto crudo stagionato. Ind. Aliment. *26*, 1158–1160.

CARERI, M., MANGIA, A., BARBIERI, G., BOLZONI, L., VIRGILI, R. and PAROLARI, G. 1993. Sensory property relationships to chemical data of Italian-type dry-cured ham. J. Food Sci. *58*, 968–972.

HABIBI-NAJAFI and LEE, B.H. 1996. Bitterness in cheese: A review. CRC Crit. Rev. Food Sci. Nutr. *36*, 397–411.

LORENZO, P. and FLORES, J. 1988. Acaros del jamón curado. Die Fleischerei 3–5.

NEY, K.H. 1979. Bitterness of peptides: amino acid composition and chain length. *In* Food Taste Chemistry. J.C. Boundreau (ed.), ACS Symp. Ser. 115, American Chemical Society, Washington, D.C.

PAROLARI, G. 1996. Achievements, needs and perspectives in dry-cured ham technology: the example of Parma ham. Food Sci. Technol. Int. *2*, 69–78.

SENTANDREU, M.A., STOEVA, S., ARISTOY, M.C., LAIB, K., VOELTER, W. and TOLDRÁ, F. 2002. Identification of taste related peptides in Spanish Serrano dry-cured hams. J. Food Sci. (Accepted for publication).

SPOTTI, E., MUTTI, P. and CAMPANINI, M. 1988. Microbiological investigation of the phenic acid defect of ham during ripening. Ind. Conserve *63*, 343–346.

TOLDRÁ, F., FLORES, J. and VOYLE, C.A. 1990. Study of the white film developed on the cut surface of vacuum-packed dry-cured ham slices. J. Food Sci. *55*, 1189–1191.

VIRGILI, R., PAROLARI, G., SCHIVAZAPPA, C., SORESI, C. and BORRI, M. 1995. Sensory and texture quality of dry-cured ham as affected by endogenous cathepsin B activity and muscle composition. J. Food Sci. *60*, 1183–1186.

CHAPTER 12

SAFETY ASPECTS

Traditionally, the dry-curing meat industry has examined the final product quality to confirm the accomplishment of the established standards, but then it is too late for any corrective action. More recently, the implementation of integrated control systems in the industry has provided a high level of product safety and full satisfaction of quality standards.

The concept of Hazard Analysis and Critical Control Points (HACCP) was originally developed in the 1960s by the Pillsbury Company when they were requested by the National Aeronautics and Space Administration (NASA) to develop a system for the assurance of the safety of foods to be consumed by astronauts in space (Bauman 1990). The objective was to avoid any illness caused mainly by the presence of bacteria, viruses or chemical contaminants. Thus, they had to control the raw material, process, personnel, environment, storage and distribution. The HACCP principles were incorporated in the low acid canned foods after its successful presentation at the 1971 U.S. National Conference on Food Protection (Buchanan 1990). Since then, HACCP has been increasingly accepted throughout the food industry, and its use has expanded rapidly in the 1990s. As an additional advantage, the HACCP system may be integrated into more general quality and safety assurance plans.

A final rule on Pathogen Reduction using HACCP systems was issued by the Food Safety and Inspection Service (FSIS) on July 25, 1996, mandating the HACCP implementation as the system of process control in all inspected meat and poultry plants. The rule addressed the problem of food-borne illness associated with meat and poultry products. This rule requires all plants to develop and write standard operating procedures for sanitation (SSOPs), to conduct microbial testing for generic *E. coli*, to develop and implement a HACCP system and to set pathogen reduction performance standards (in the case of slaughterhouses and plants producing raw ground products) for *Salmonella*. FSIS determined generic models for each of nine processing categories and made them available for the assistance of establishments in the preparation of plant-specific HACCP plans. All the details and the generic models can be obtained in FSIS updated revisions of the generic models such as the generic HACCP model for raw, not ground meat and poultry products (USDA 1999a) or the generic HACCP model for not heat treated, shelf stable meat and poultry products (USDA 1999b).

Since January 26, 1998, large plants (those with more than 500 employees) have been required to meet the specifications of the Pathogen Reduction, HACCP Systems final rule. All small plants (those having between 10 and 500 employees) were required to implement HACCP since January 25, 1999. Finally, the implementation of the rule in very small establishments (less than 10 employees or annual sales of less than \$2.5 million) was made effective by January 25, 2000. The small and very small plants are responsible for most of the production of dry-cured meat products.

The HACCP system involves several steps that, if not properly handled, may result in the failure of the full plan. These steps are as follows: (1) identification of hazards associated with the production, distribution, sale and consumption of the product (a hazard is considered as anything dangerous to human health and is reasonably likely to exist in the food); (2) determination of critical control points (CCP) where loss of control would result in an unacceptable risk to the consumer: CCP1 when the risk is fully eliminated and CCP2 when it is minimized; (3) establishment of critical limits assuring that a hazard is under control; (4) establishment of a monitoring system to ensure that each CCP is under control and maintenance of full records for management, audits, trend analysis and scrutinizing by inspectors; (5) protocols for CCP deviations detailing the corrective action that should be taken to get the CCP under control; (6) establishment of procedures for verification of the system (this is necessary for confirming the effectiveness of control measures by collecting supplementary information, such as microbiological tests, which are used to assess the effectiveness of the HACCP system, and (7) establishment of effective (as accurate and user-friendly as possible) record-keeping systems that will facilitate the access to anybody involved in the process as well as auditors.

The HACCP must be done for each specific process and product, taking into account the raw materials, equipment, operating procedures, packaging, storage, distribution and intended use conditions (Mayes and Baird-Parker 1992). Implementation of a HACCP system in the dry-curing industry requires a multidisciplinary team for correct information and assessment.

GENERIC HACCP MODEL FOR DRY-CURED HAMS

The FSIS generic HACCP model for raw (not ground) meat and poultry products has important potential benefits assisting the establishment in applying the seven HACCP principles to meet their ham processing operations, including a number of forms that can be used to record various types of data and information.

The company's HACCP team must describe the product either as a process flow diagram indicating the stages that are critical for the safety of the product (Fig. 12.1) or as a written form indicating the process category, type of product, common name, mode of usage, type of package, shelf-life, labelling instructions and necessary controls during distribution. This is helpful for completing the hazard analysis.

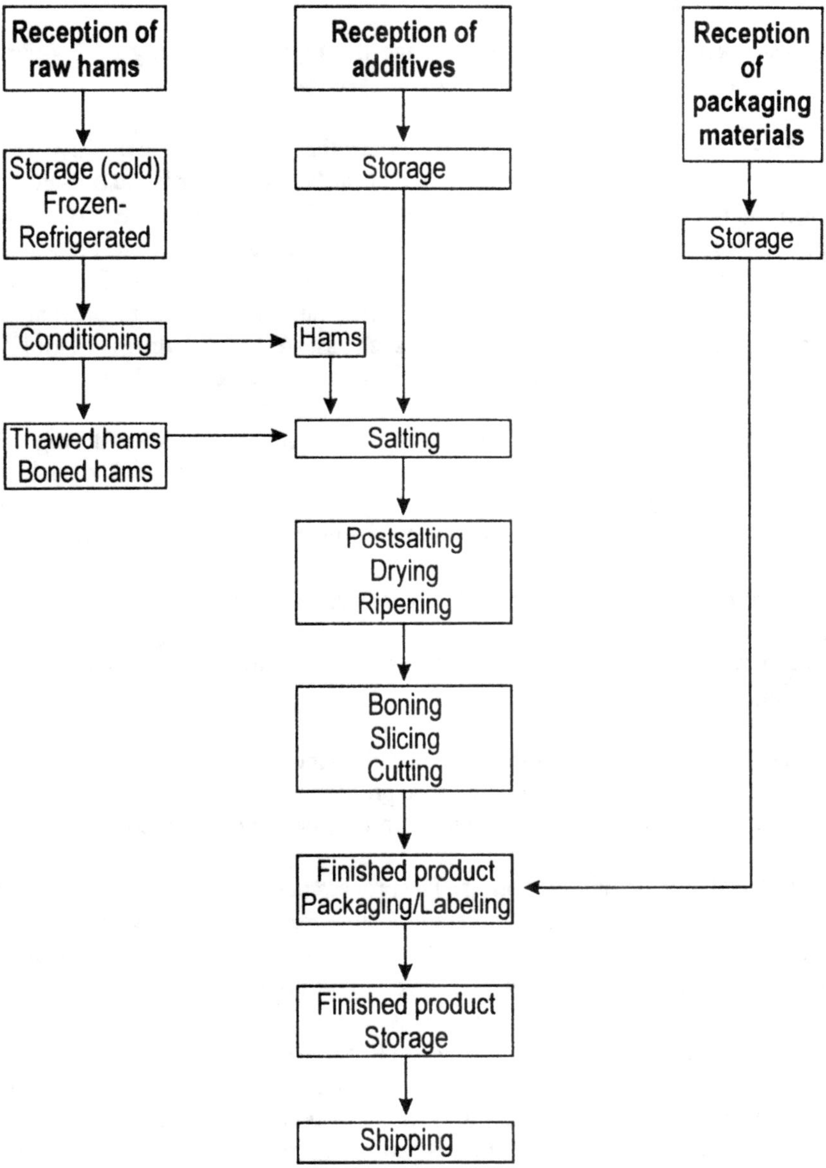

FIG. 12.1. PROCESS FLOW DIAGRAM INDICATING THE STAGES CRITICAL FOR THE SAFETY OF DRY-CURED HAM

Identification of Hazards

The identified hazards for each stage during the processing of dry-cured ham are briefly described in the following section.

Raw Materials. All products and services that may affect the quality of raw materials must be considered. Among the hazards to consider is the presence of dangerous microorganisms in the ham (e.g., *Cl. botulinum* or *S. aureus*), in the ingredients or in the water as well as other microorganisms deleterious to the product (e.g., anaerobic Micrococcaceae or Lactic acid bacteria). The thermal history of the hams is important. DFD hams must be avoided to prevent the microbial growth. Chemical hazards such as veterinary drugs (antibiotic residues) or contaminants (pesticides or dioxins) must be considered also.

Storage. The main hazard in storage is pathogen proliferation during cold storage. Correct conditions (time/temperature) must be maintained during the storage of raw meat, either under refrigeration or freezing. Hygienic conditions are likewise important for handling and storing the ingredients and additives.

Conditioning. This includes, if necessary, thawing, boning, washing and classification of hams. The main risk is the outgrowth of microbial flora that either already exists in the ham or has been acquired from an operator's handling.

Salting. The main hazard in salting is the use of contaminated salt. Insufficient salt penetration can also cause problems. Additionally, inappropriate conditions (like temperature) or hygiene must also be considered.

Post-salting/Drying/Ripening. The process parameters that may influence ham safety and/or stability are temperature, time, relative humidity and air speed rate. These parameters must be carefully controlled. For example, an incorrect drying with high water activity may favor microorganism growth.

Packaging. Hams may be boned and optionally sliced or cut into small pieces. This is one of the points prone to contamination since handling presents the opportunity to accidentally introduce pathogenic microorganisms or other contaminants. The slices and pieces are vacuum-packaged or kept under modified atmosphere, so the product container integrity and sealing operations must also be controlled.

Finished Product Storage. The main hazard in finished product stage is pathogen proliferation in the finished product. Correct conditions (time/temperature) for storage must be kept as well as hygienic conditions for handling and storing the hams.

Distribution and Sale. The risks with distribution and sale are associated with hygienic conditions and correct handling in distribution to centers and retail outlets.

Customer Practices. Customer practices involve handling in the kitchen by the consumer or professional cook like slicing, cooking, serving, etc.

Control Measures

The growth of microorganisms can be minimized by low temperatures (during reception, storage, salting and post-salting), the addition of preservatives (salt and nitrates added in the curing salt) and moisture reduction (during drying). Shipments must be checked for pathogen certification. Correct design and operation of the drying/ripening chambers is essential, especially the monitoring of temperature, time and relative humidity as well as the calibration of the sensors. Residues of veterinary drugs can be controlled by allowing a certain time between application and slaughter or online detection by the application of rapid test kits at the slaughterhouse. Handling procedures must be sanitary, contact surfaces must be clean and the environment should not introduce any foreign material or microorganism. The packaging process must be done in an equally sanitary manner.

Critical Control Points

Several CCPs that will minimize the risk or prevent the hazard from occurring are identified at each stage, and control procedures (critical limits, tolerances and monitoring systems) are established for each CCP. For example, the raw hams must be inspected during their reception for any sign of defect or alteration, and tests may be conducted for checking the organoleptic conditions or presence of microbial or chemical residues. The critical limits for refrigerated or frozen hams must be below 7C or -12C, respectively. The pH should not exceed 5.8 at the beginning of the process since higher values would constitute a microbiological risk during later stages.

The safety of the hams can be verified by microbiological analysis. Ingredients and tap water must conform to microbiological requirements. There is a need for rapid physical/chemical tests and visual inspection since monitoring methods must be rapid to be effective. The corrective action is the rejection of suspected lots of hams that do not accomplish the required conditions until further examination demonstrates that the product is safe. Further action must be taken to prevent it from happening again. Full records must be kept for all monitoring data.

The effectiveness of the system also depends on the availability of qualified personnel, the reliability of the methodology and the accuracy of the control records. Verification that the system is working reassures the producer that the application of HACCP is effective for the production of safe dry-cured hams. Finally, all the information, including data used in the study and all decisions reached, must be recorded and readily accessible.

GENERIC HACCP MODEL FOR DRY-FERMENTED SAUSAGES

The FSIS generic HACCP model for not heat treated, shelf stable meat and poultry products has important potential benefits assisting the establishment in applying the seven HACCP principles. It also includes a number of forms for recording data and information. As in the case with hams, the company's HACCP team must describe the product either as a process flow diagram indicating the stages that are critical for the safety of the product (Fig. 12.2) or as a written form indicating the process category, type of product, common name, mode of usage, type of package, shelf life, labeling instructions and necessary controls during distribution.

Identification of Hazards

Lactic acid bacteria play a critical role in safety and preservation of dry-fermented sausages by fermenting carbohydrates to organic acids (mainly lactic acid) and thus producing a pH drop which inhibits the acid-sensitive spoilage microorganisms. Additionally, the competition for essential nutrients also constitutes a selective process in favor of Lactic acid bacteria. The production, even in small amounts, of other substances like free fatty acids, ammonia, ethanol, hydrogen peroxide, carbon dioxide and bacteriocins may also contribute to a protective effect against a wide spectrum of microorganisms (De Vuyst and Vandamme 1994). Bacteriocins, which are produced by strains of Lactic acid bacteria belonging to the genera *Lactobacillus, Pediococcus, Leuconostoc, Carnobacterium* and *Enterococcus* (Aymerich *et al.* 1998), have increased interest in recent years. These compounds are biologically active proteins or protein complexes displaying a bactericidal action against other related microorganisms at micromolar concentrations (De Vuyst and Vandamme 1994; Leroy and Vuyst 1999). Some bacteriocins are only active against bacteria belonging to the same genus, while others are active against other bacteria

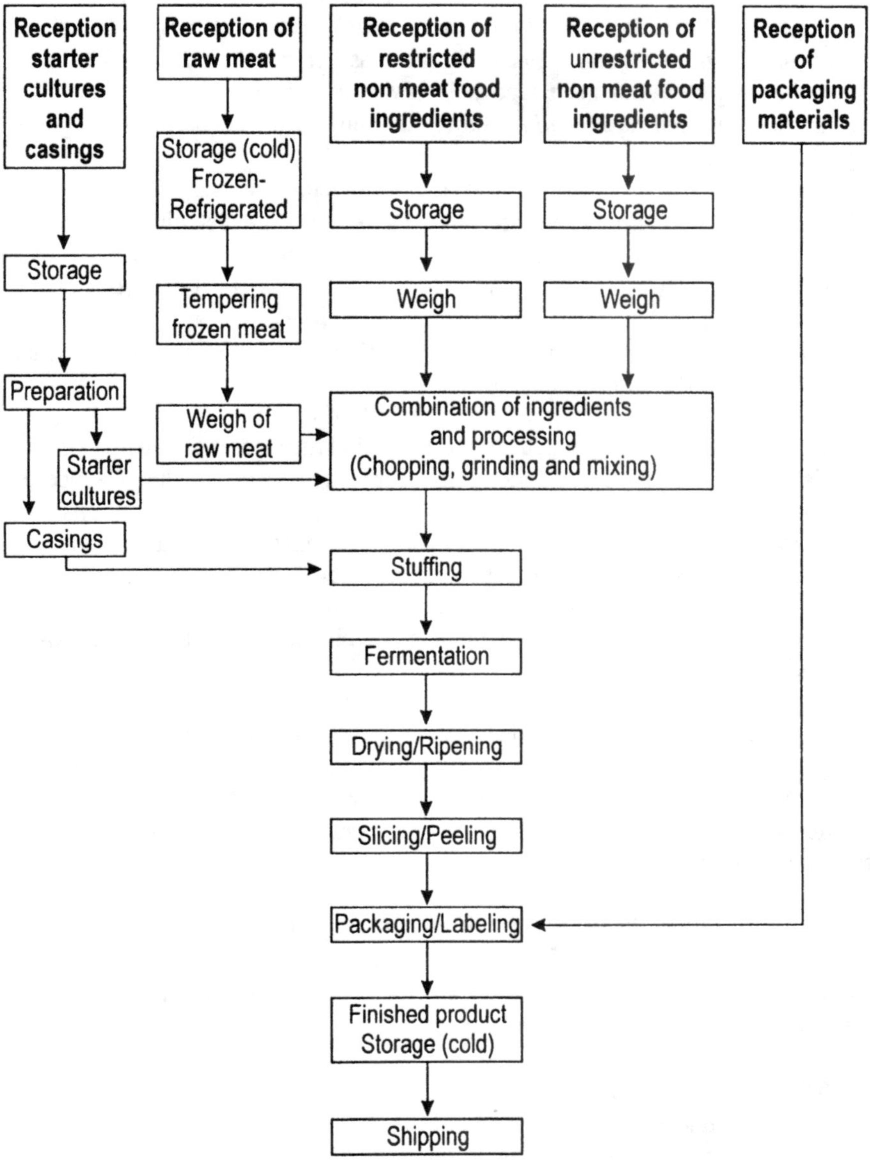

FIG. 12.2. PROCESS FLOW DIAGRAM INDICATING THE STAGES CRITICAL FOR THE SAFETY OF DRY-FERMENTED SAUSAGES

genera such as *Listeria monocytogenes, Staphylococcus aureus, Clostridium perfringens, Clostridium botulinum* and *Brochothrix thermosphacta*. The inhibition of gram-negative bacteria such as *Aeromonas hydrophila* and *Pseudomonas putida* has been reported only in a few cases, while others like *Escherichia coli* and *Salmonella* require the addition of chelating-like agents (Aymerich *et al.* 1998).

The presence of lactobacilli is not the only factor necessary for producing safe and stable sausages. There are other factors such as the initial anaerobic conditions, the addition of salt and nitrite and the progressive reduction in water activity, which constitutes successive barriers, known as the hurdle effect (Leistner 1992), to the growth of undesirable microorganisms.

The identified hazards for each stage during the processing of dry-cured sausages are briefly described below.

Raw Materials. All products and services that may affect the quality of raw materials must be considered. Among the hazards to consider is the presence of dangerous microorganisms in the meats (e.g., *Cl. Botulinum* or *S. aureus*), in the ingredients or in the water as well as other microorganisms potentially deleterious to the product (e.g., anaerobic Micrococcaceae or Lactic acid bacteria). The thermal history of the meats is important. Chemical hazards like veterinary drugs (antibiotic residues) or contaminants (pesticides or dioxins) must be considered also.

Storage. The main hazard in storage is pathogen proliferation during cold storage. The correct conditions (time/temperature) must be maintained for storing the raw meat either under refrigeration or freezing. Of equal importance are the hygienic conditions for handling and storing the ingredients, microbial starters and additives.

Conditioning. This includes, if necessary, thawing, boning, cutting, washing and classification of the meats. The main risk is the outgrowth of microbial flora that is either already existing in the meats and fats or has been acquired from the operator's handling.

Mixing. The main hazards in mixing are the use of contaminated additives, incorrect formulation and metal contamination during mechanical processing. Additionally, inappropriate handling, environmental conditions (like temperature) or hygiene must be also considered.

Stuffing. The main hazard in stuffing is due to contamination by non-hygienic casings. For instance, if natural casings are used, they must be thoroughly desalted and decontaminated. Another hazard is due to the accumulation of bubbles inside the sausage or holes in the casing if lower or higher pressures are used than are indicated during stuffing.

Fermentation/Drying/Ripening. The main hazards in these steps are pathogen proliferation during fermentation and drying. So, the process parameters like fermentation temperature, time, relative humidity and air velocity have to be carefully controlled since they have a direct influence on the sausage safety, stability and development of organoleptic characteristics. An inadequate development of fermentation may result in product alteration such as inadequate pH drop, growth of undesirable microorganisms, sensory defects, etc. pH and a_w must also be controlled because they will limit the microflora growing inside the sausages.

Packaging. The slicing is one of the processing points prone to contamination since handling presents the opportunity to accidentally introduce pathogenic microorganisms such as *Listeria monocytogenes* or other contaminants. The slices and pieces are vacuum-packaged or kept under modified atmosphere. Therefore, the product container integrity and sealing operations must be also controlled.

Finished Product Storage. The main hazard in finished product storage is that the correct conditions (time/temperature) for storage as well as hygienic conditions for handling and storing the sausages must be maintained. The microflora on the outer surface of the sausage must be controlled.

Distribution and Sale. The risks of distribution and sale are associated with hygienic conditions and correct handling in distribution centers and retail outlets.

Customer Practices. Customer practices involve the handling in the kitchen by the consumer or professional cook, like slicing, cooking, serving, etc.

Control Measures

The growth of undesirable microorganisms can be prevented or limited by the use of low temperatures (during reception, storage and mixing), the addition of preservatives (salt and nitrates/nitrites added in the curing salt) and/or the competition with microbial starters. Correct design and operation of the drying/ripening chambers is essential, especially the monitoring of temperature, time and relative humidity as well as the calibration of plant sensors. As in the case of hams, residues of veterinary drugs can be controlled by allowing a certain time between application and slaughter or online detection by the application of rapid test kits at the slaughterhouse. Handling procedures must be sanitary, contact surfaces must be clean and the environment should not introduce any foreign material or microorganism. The packaging process must be done in an equally sanitary manner.

Critical Control Points

Several CCPs were proposed by Leistner (1985), taking into account the specific microbiological risks of dry sausages and the important control points. Some of them are identical to those previously described for hams. This is the case for raw materials where a control of the thermal history and colony counts of the raw meats and fats is very important. The safety of the meats and fats can be verified by microbiological analysis. Ingredients, additives and tap water must also comply with microbiological requirements. As in the case of hams, monitoring methods must be rapid to be effective. Important critical control points are the reception and proper cold storage of raw materials, the mixing stage (addition of salt, sugar, nitrate, nitrite, spices and starter cultures to the sausage mix) as well as the casings, processing conditions (such as fermentation temperature, relative humidity and air velocity), correct pH reached after the fermentation process, proper a_w reached after drying, proper microflora in the interior of sausages as well as the desirable and undesirable microorganisms on the surface, metal detectors prior to packaging and proper sanitizer used at slicing.

The HACCP team must establish critical limits for each CCP, monitoring procedures and their respective frequencies as well as planning corrective actions to be followed in response to a deviation from a critical limit. As in the case of hams, the system must be verified periodically to reassure the producer that the application of HACCP is effective for the production of safe dry-fermented sausages and all the information (including CCPs definition, critical limits, descriptions of control procedures, modifications, monitoring and verification) is recorded and readily accessible.

Dry-cured meats have provided a healthy, nutritional and appetizing dietary components for centuries. Now, with modern and easily used programs of control, both quality and safety can be further assured.

REFERENCES

AYMERICH, M.T., HUGAS, M. and MONFORT, J.M. 1998. Review: Bacteriocinogenic lactic acid bacteria associated with meat products. Food Sci. Technol. Int. *4*, 141–158.

BAUMAN, H. 1990. HACCP: Concept, Development and Application. Food Technol. *44*, 156–165.

BUCHANAN, R.L. 1990. HACCP: A re-emerging approach to food safety. Trends Food Sci. Technol. *1*, 104–106.

DE VUYST, L. and VANDAMME, E.J. 1994. Antimicrobial potential of lactic acid bacteria. *In* Bacteriocins of Lactic Acid Bacteria. L. De Vuyst and E.J. Vandamme (eds.), p. 91. Chapman & Hall, Oxford, U.K.

LEISTNER, F. 1985. Stable and safe fermented sausages world-wide. *In* Fermented Meats. G. Campbell-Platt and P.E. Cook (eds.), pp. 160. Blackie Academic & Professional, London, U.K.

LEISTNER, F. 1992. Meat fermentation as an integrated process. *In* New Technologies for Meat and Meat Products. J.M. Smulders, F. Toldrá, J. Flores and M. Prieto (eds.), pp. 1–19. Audet, Nijmegen, The Netherlands.

LEROY, F. and DE VUYST, L. 1999. Temperature and pH conditions that prevail during fermentation of sausages are optimal for production of the antilisterial bacteriocin sakacin K. Appl. Environ. Microbiol. *6*, 947–981.

MAYES, T. and BAIRD-PARKER, A.C. 1992. Hazard analysis studies. *In* New Technologies for Meat and Meat Products. J.M. Smulders, F. Toldrá, J. Flores and M. Prieto (eds.), pp. 263–272. Audet, Nijmegen, The Netherlands.

USDA 1999a. Generic HACCP Model for Raw, Not Ground Meat and Poultry Products, FSIS, USDA, HACCP-4, www.fsis.usda.gov/index.htm.

USDA 1999b. Generic HACCP Model for Not Heat Treated, Shelf Stable Meat and Poultry Products, FSIS, USDA, HACCP-15, www.fsis.usda.gov/index.htm.

CHAPTER 13

ECONOMIC AND INTERNATIONAL ASPECTS

New trends in food consumption are arising with the global market. Industry and consumers change in their opinions of tastes and flavors and adopt a worldwide attitude toward food. Because the current trend is smaller families or single person households, less and less time is devoted to preparing meals in the kitchen. Therefore, there is a need for convenience foods even though they are more expensive.

The changing patterns in eating and the cosmopolitan tastes of the consumers have driven the research for new products or for the improvement of those products already existing. For instance, there is the availability of sliced dry-cured ham or sausages, which are ready for consumption with no further need for cooking. Since there is an evident movement toward snacking, dry-cured meat products become an interesting convenience food because they can be stored for several weeks under refrigeration (with no need for immediate consumption), they are already cut into slices (with no need to be cut), they have a high nutritive value and they are ready to eat (with no need for cooking).

This trend is reflected in Europe where Northern European countries are increasing their consumption of dry-cured meats and the industries are getting more and more interested in the processing of products with longer ripening times.

TRENDS IN WORLD MEAT PRODUCT CONSUMPTION AND PRODUCTION

It is rather difficult to obtain accurate figures on the production and consumption of dry-cured meat products that are separate from the total of processed meat products. In this regard, only estimates of the size of the market can be given.

European Union

Pork products represent 48% of total meat production in the European Union (E.U.), with Germany, France and Spain being the major producers. However, the major exporter is Denmark. The production of meat products during 1997 in the European Union is shown in Table 13.1. Main producers of dry-cured hams and shoulders are Italy, Spain and Germany, which altogether account for approximately 80% of the estimated production in the E.U. (Fisher and Palmer 1995). Main producers of fermented sausages are Germany, Spain

TABLE 13.1.
PRODUCTION (IN TONNES) OF DRY-CURED MEAT PRODUCTS IN THE EUROPEAN UNION (1997)

Country	Dry-cured hams and shoulders	% of total	Other dry or smoked meat products	% of total	Cooked and Dry-fermented sausages	% of total
Germany	164,169	21	168,395	23	1,183,249	40
United Kingdom	67,611	9	190,922	26	326,691	11
France	14,002	2	51,205	7	331,049	11
Italy	333,675	43	30,738	4	36,471	1
Spain	127,247	17	26,550	4	393,483	13
Denmark	1,507	<1	123,535	17	99,859	3
The Netherlands	3,741	<1	49,428	7	122,684	4
Belgium	17,427	2	15,919	2	107,266	4
Sweden	6,282	1	26,070	4	121,244	4
Austria	7,170	1	13,573	2	134,315	5
Ireland	15,796	2	32,764	4	20,832	1
Portugal	4,893	1	1,716	<1	20,915	1
Finland	5,271	1	1,716	<1	-	-
Greece	1,797	<1	415	<1	30,472	1
TOTAL	770,587	100	732,603	100	2,928,532	100

(Source: Liaison Centre for the Meat Processing Industry in the EU, CLITRAVI)

and France. In 1999, the total amount of exported dry-cured hams by the E.U. was 14,000 tons, and dry-fermented sausages was 11,600 tons. Imports of meat products in the E.U. by country of origin is given in Table 13.2. The amounts of imports and exports are relatively low in comparison to respective productions. Inside the E.U., main exporters are similar to major producers. Main exports and imports are shown in Table 13.3.

TABLE 13.2.
IMPORTS OF MEAT PRODUCTS IN THE EU BY COUNTRY OF ORIGIN (1997)

Country	Dry-cured meats (tonnes)	Dry sausages (tonnes)
Hungary	80	5,344
USA	57	10
Rep. Czech	0.2	4
Poland	7	680
Canada	1	34

(Adapted from Pozo 1998)

TABLE 13.3.
MAIN EXPORTS AND IMPORTS OF DRY-CURED HAMS BY COUNTRIES (1997)

Country	Export (tonnes)	Country	Import (tonnes)
Spain	32,843	France	15,000
Italy	9,254	Germany	10,000
France	7,084	Belgium	4,800
Germany	6,500	Austria	4,000
Belgium	2,451	Argentina	3,800

(Adapted from Macías 1998)

The Spanish production of dry-cured hams and shoulders was as high as 200,510 tons, representing more than 30 million pieces in 2000 (Cruz and Barreiro 2001). The production of dry-fermented sausages in 2000 was also very high, 169,999 tons. Italian Parma ham Consortium produced 8,064,000 hams in 1997 (Utini 1998). An example of exports from Italy and Spain to other countries is shown in Table 13.4. The main markets for boned dry-cured Parma hams are France, Germany and the U.S. In the case of Spanish Serrano hams, exports amounted to 12,020 tonnes and main destinations are France, Germany, Argentina, Portugal and, on a smaller scale, Belgium, the U.S., the Netherlands and Russia. Today, the exports are still low but are increasing steadily, and the perspectives are very good.

TABLE 13.4.
EXPORTS OF DRY-CURED HAM FROM ITALY (8,620 TONNES IN 1997) AND SPAIN (12,020 TONNES IN 1999) TO OTHER COUNTRIES.

Country	Exports (%)	Country	Exports (%)
France	36.7	France	28.3
Germany	25.6	Germany	20.7
USA	11.2	Argentina	16.0
Belgium	6.3	Portugal	15.3
Switzerland	5.1	Belgium	2.6
United Kingdom	4.2	Others	17.1
Japan	3		
Other[a]	9		

[a]Argentina and Andes Pact countries; [b]USA, The Netherlands, Russia
(Adapted from Utini 1998, and Pozo 2000)

Nearly 50% and 60% of all pig meat consumed in Spain and Italy, respectively, are in the form of processed products. The highest consumption of dry-cured hams and shoulders is found in Spain, around 4.4 Kg per person per year, followed by Italy with 2.4 Kg per person per year, France with 0.9 Kg per person per year, Belgium with 0.6 Kg per person per year and Germany with 0.4 Kg per person per year (Macías 1998).

U.S. and Other Countries

The import and commercialization of dry-cured meat products must be controlled by regulatory agencies. One of the most important points of control is in the inactivation of different viruses in the cured products so that there is no risk for the country's own livestock. Several studies have been carried out on dry-cured meat products (Mebus *et al.* 1997). Recent studies have been performed on the inactivation of viruses in dry-cured hams such as the Italian Parma ham (McKercher *et al.* 1985, 1987) and the Spanish Serrano and Iberian hams (Mebus *et al.* 1993a, b). In this last case, an agreement between the USDA, the Spanish Ministry of Agriculture, Food and Fisheries and the Associations of Meat Industries was signed in 1987 to carry out a research project that would recognize if the dry-curing process for hams, shoulders and loins eliminated the agents causing porcine epizooites, which were of main economic importance. The viruses highlighted in the inactivation study were: African swine fever virus, classical swine fever virus, foot and mouth disease virus and swine vesicular disease virus.

The results of this research confirmed the theory (Mebus *et al.* 1993a, b) and allowed for the publication, on November 3, 1994, of the permission for importation of Spanish dry-cured hams processed in determined conditions (Federal Register 1994). Recently, the status of Spain, because of swine vesicular disease, has been modified to the consideration of a country free of this disease (Federal Register 1995). It has contributed in reducing the minimum period of processing for both Serrano and Iberian dry-cured hams to 190 and 365 days, respectively.

The U.S. Department of Agriculture (USDA) published a detailed regulation about specific time-temperature combinations that are acceptable to assure complete inactivation of the *Trichinella spiralis* (Marriot *et al.* 1992). Main imports of meat products in the U.S. during 1999 came from Canada (1,326 tonnes) and the E.U. (1,072 tonnes). The amounts are very low, probably due to the required process of homologation of the production plant, slaughterhouse and cutting plant that is complex and restricts the number of companies able to export. The exports to the European Union were also very low.

Argentina is the main destination for sliced, boned dry-cured ham, followed by the U.S. outside of Europe. China produces about one million dry-cured hams in several provinces in the south of the country. Some of these hams are exported to Hong Kong, Taiwan, Singapore and other Southeast countries (Zhu 1998).

REFERENCES

CRUZ, J. and BARREIRO, D. 2001. Situación actual de la industria cárnica en España. Eurocarne *95*, 23–31.

FISHER, S. and PALMER, M. 1995. Fermented meat production and consumption in the European Union. *In* Fermented Meats. G. Campbell-Platt and P.E. Cook (eds.), pp. 217–233. Blackie Academic & Professional, London.

MACÍAS, F. 1998. Dry-cured ham in Europe. Presented at the Symp. on Dry-cured ham: Technology and consumption analysis. September 1, Barcelona, Spain.

McKERCHER, P.D., BLACKWELL, J.H., MURPHY, R., CALLIS, J.J., PANINA, G.F., CIVARDI, A., BUGNETTI, M., DeSIMONE, F. and SCATOZZA, F. 1985. Survival of swine vesicular disease virus in Prosciutto di Pama (Parma ham). Can. Inst. Sci. Food Technol. J. *18*, 163–167.

McKERCHER, P.D., BLACKWELL, J.H., MURPHY, R., CALLIS, J.J., PANINA, G.F., CIVARDI, A., BUGNETTI, M., DeSIMONE, F. and SCATOZZA, F. 1987. Survival of viruses in Prosciutto di Pama (Parma ham). Can. Inst. Sci. Food Technol. J. *20*, 267–272.

MEBUS, C.A., HOUSE, C., RUIZ-GONZALVO, F., PINEDA, J.M., TAPIADOR, J., PIRE, J.J., BERGADA, J., YEDLOUTSCHNING, R.J., SAHU, S., BECERA, V. and SÁNCHEZ-VIZCAÍNO, J.M. 1993a. Survival of Foot-and-mouth disease, African swine fever and Hog cholera viruses in Spanish Serrano cured hams and Iberian cured hams, shoulders and loins. Food Microbiol. *10*, 133–144.

MEBUS, C.A., HOUSE, C., RUIZ-GONZALVO, F., PINEDA, J.M., TAPIADOR, J., PIRE, J.J., BERGADA, J., YEDLOUTSCHNING, R.J., SAHU, S., BECERA, V. and SÁNCHEZ-VIZCAÍNO, J.M. 1993b. Survival of Swine vesicular disease virus in Spanish Serrano cured hams and Iberian cured hams, shoulders and loins. Food Microbiol. *10*, 263–268.

MEBUS, C.A., ARIAS, M., PINEDA, J.M., TAPIADOR, J., HOUSE, C. and SÁNCHEZ-VIZCAÍNO, J.M. 1997. Survival of several porcine viruses in different Spanish dry-cured meat products. Food Chem. *59*, 555–559.

POZO, M. 1998. La industria cárnica en Europa. Eurocarne *59*, 37–46.

POZO, M. 2000. Exportación del jamón curado. Eurocarne *89*, 69–71.

UTINI, D. 1998. Parma ham. Presented at the Symp. on Dry-cured ham: Technology and consumption analysis. September 1, Barcelona, Spain.

ZHU, S. 1998. Dry-cured ham in China. Presented at the Symp. on Dry-cured ham: Technology and consumption analysis. September 1, Barcelona, Spain.

INDEX

www.ingramcontent.com/pod-product-compliance
Lightning Source LLC
LaVergne TN
LVHW011754200926
848484LV00003B/706
9780917678547